SIMPLICIUS
On Aristotle Physics 3

SIMPLICIUS
On Aristotle Physics 3

Translated by J.O. Urmson

Notes by Peter Lautner

Duckworth

First published in 2002 by
Gerald Duckworth & Co. Ltd.
61 Frith Street
London W1D 3JL
Tel: 020 7434 4242
Fax: 020 7434 4420
email: enquiries@duckworth-publishers.co.uk
www.ducknet.co.uk

Introduction © 2002 by Richard Sorabji
Translation © 2002 by J.O. Urmson
Notes © 2002 by Peter Lautner

All rights reserved. No part of this publication may be reproduced, stored in a retrieval system, or transmitted, in any form or by any means, electronic, mechanical, photocopying, recording or otherwise, without the prior permission of the publisher.

A catalogue record for this book is available from the British Library

ISBN 0 7156 3067 9

Acknowledgments

The present translations have been made possible by generous and imaginative funding from the following sources: the National Endowment for the Humanities, Division of Research Programs, an independent federal agency of the USA; the Leverhulme Trust; the British Academy; the Jowett Copyright Trustees; the Royal Society (UK); Centro Internazionale A. Beltrame di Storia dello Spazio e del Tempo (Padua); Mario Mignucci; Liverpool University; the Leventis Foundation; the Arts and Humanities Research Board of the British Academy; the Esmée Fairbairn Charitable Trust; the Henry Brown Trust; Mr and Mrs N. Egon; the Netherlands Organisation for Scientific Research (NWO/GW). The editor wishes to thank Keimpe Algra, Rachel Barney, Charles Brittain, Jan Opsomer, Gerd Van Riel and Christian Wildberg for their comments, and Eleni Volonaki and Han Baltussen for preparing the volume for press.

Typeset by Ray Davies
Printed in Great Britain by
Bookcraft (Bath) Limited, Midsomer Norton, Somerset

Contents

Introduction	1
Translator's Note	5
Textual Emendations	7
Translation	11
Notes	145
English-Greek Glossary	169
Greek-English Index	173
Subject Index	197

Introduction
Richard Sorabji

Aristotle's *Physics* Book 3 covers two main subjects, the definition of change and the finitude of the universe. The *Physics* is about nature, and change enters into the very definition of nature as an internal source of change. Change receives two definitions in chapters 1 and 2, as involving the actualization of the potential (201a10-11), or of the changeable (202a7-8). Alexander is reported (Simplicius *in Phys.* 436,26-32) as holding, and Philoponus agrees (*in Phys.* 367,6-369,1), that the second definition is designed to disqualify change in relations from being genuine change.

Relational change

Immediately before his first definition, at 200b33-201a9, Aristotle leaves a rather confusing impression as to whether there is change in respect of relation, or only in respect of place, quantity, quality and substance. Having started by saying that change (*kinêsis, metabolê*) is always (*aei*) in these four respects, he finishes by saying that there are as many kinds of change as there are categories of being, which would suggest ten kinds. The ambiguity is resolved in *Physics* 5, where he says that in respect of relatives there is no change, except accidental (*kata sumbêkos*), which he has agreed to dismiss (*apheisthô*, 224b7), because it is possible that *without a thing's changing at all*, a different relational property becomes true of it (*Phys.* 5.2, 225b11-13 = *Metaph.* 11.12, 1068a11-13; similarly 7.3, 246b11-12; *Metaph.* 14.1, 1088a30-5). The idea that one of the two relatives 'does not change at all', when e.g. Socrates comes to be shorter than the growing Theaetetus, is taken from Plato *Theaetetus* 155B-C, and is further endorsed by the Stoics, ap. Simplicium *in Cat.* 166,17-29; 172,1-5. The idea of relational change has been reintroduced into modern discussions under the title of 'Cambridge change' by Peter Geach.

Simplicius is worried at Aristotle's restriction. He would rather that Aristotle had made a terminological distinction, allowing 'transformation' (*metabolê*), as the wider term, to extend beyond the four categories of change (*kinêsis*) *in Phys.* 861,5-20. Moreover he cites Aristotle's successor Theophrastus (*in Cat.* 435,28-31; *in Phys.* 412,31-413,9) and Alexander (*in Phys.* 409,12-32) as allowing change in the other categories. But Theophrastus, he thinks, goes too far, in not distinguishing change of place, quantity and quality, as affecting the disposition (*diathesis*) of a thing, from changes which merely affect its relationship (*skhesis*), a distinction suggested by Aristotle's pupil Eudemus, *in Phys.* 861,5-26; 859,16-27.

Activity of agent and patient identical

In *Physics* 3.3, Aristotle maintains that the activity of the teacher is located in the learner and in a way identical with the activity of the learner, though not the same in definition. Rather, if you are counting how many activities are going on, there is only one to be counted. This enables Aristotle in *Physics* 8, as Simplicius observes *in Phys.* 442,18, to locate the activity of the divine unmoved mover in the universe which he moves, and so to accommodate no motion within himself. Philoponus *in Phys.* 385,4ff. refers to Aristotle's principle, in order to support his own impetus theory, according to which the impetus imparted by a thrower comes to be located *inside* the projectile.

The doctrine has another major importance for Neoplatonism, for it is the basis of Aristotle's view in *On the Soul* 3.2, 3.7 and 3.8, that the activity of the perceptible or intelligible is in a way identical with the activity of perception or intellect. It is central to Neoplatonism that Intellect can be identical with its intelligible objects, the Platonic Forms, although this identity allows that the activity of the intelligibles, like that of the teacher, acts as agent and so has a certain priority. The identity also means that, in being aware of its objects, Intellect is in a way aware of itself. It further gives the human intellect the opportunity of being united with the Forms, while at the same time sowing the seeds of the Averroist problem about how disembodied intellects, if united with Forms, are still distinct from each other.

Universe spatially finite

In *Physics* 3.4-8, Aristotle analyses infinity, and concludes that the universe is spatially finite. He gives an account of infinity still propounded by modern school teachers, according to which there is never a more than finite number of anything, but to talk of infinity is to say that however large a finite number of something you have, you can always have a larger finite number. Infinity is thus an ever expandable finitude, just as it is in modern talk of approaching a limit, or getting as close as you like. This helps to make it seem less frightening. One of Aristotle's objections to a more than finite number is that its parts would also be more than finite, *Phys.* 3.5 204a20-9. That is something whose acceptability was not explained in the West until the fourteenth century, although it has been pointed out to me that it was known in the thirteenth century to

Introduction

Grosseteste, I presume from an Arabic source, and I would tentatively think of al-Haytham.

In defending Aristotle's view that the universe is spatially finite, Simplicius has to answer the objection of Plato's Pythagorean friend Archytas (*in Phys.* 467,26-32), 'What happens at the edge? Can I stretch out my hand or stick, or not?' The objection had been elaborated by Eudemus, and had been answered by Alexander, *Quaestiones* 3.12, 106,35-107,4. There is nothing rather than empty space beyond the furthest stars, and one cannot stretch into nothing, nor even want to. That there is no place at all for stretching would follow from Aristotle's definition of place in terms of a thing's *surroundings*. Beyond the furthest stars, there are no surroundings. Simplicius slightly alters Alexander's solution when he repeats it, *in Phys.* 467,35-468,3; cf. *in Cael.* 285,21-7. A man is not *prevented* from stretching by nothingness. Rather, nothingness neither repels nor accommodates a hand.

Simplicius does not at 516,3-38, fully bring out the reply sketched by Aristotle *Physics* 3.8, 208a11-20, and elaborated by Alexander *Quaestiones* 3.12, 104,24-9, which tackles the objection that if we try to think of the universe as limited, we have to think of it, self-defeatingly, as limited by something *further out*. Their answer is that 'limited' has a different logic from 'touched'. It does not equally imply an agent doing the limiting. Alexander illustrates the point by offering three sufficient conditions for a whole being limited, none of which implies a limiting agent outside. I have interpreted these three conditions in *Matter, Space and Motion*, ch. 8, at 135-8, as follows. First, a whole can have a limit, if it has a limit in Aristotle's sense of a first rim outside which you cannot take anything. Secondly, in Alexander's view, a thing will be limited, if it can be divided into an equal number of segments. Thirdly, a whole can have a limit if it consists of a limited number of parts of limited size, where this last reference to the limitedness of the parts can be understood in the opponent's way, as being limited by other parts. What is the meaning of 'limited number' in Alexander's suggestion? Simplicius reports, 516,29-38, Alexander's remark that a number can be limited (*peperasthai*) without possessing a limit (*peras*) at all. Simplicius thinks this impossible, but surely Alexander is right.

Universe temporally finite?

Philoponus was to complain of an asymmetry in Aristotle. His universe ought to be temporally, as well as spatially, finite, and that would refute pagan Aristotelians and Neoplatonists and vindicate the Christian view that the universe had a beginning. Without a beginning, the universe will have finished going right through a more than finite number of years and that will be only a fraction of the more than finite number of days. Finishing an infinity and infinite fractions had both been ruled out by

Aristotle. Simplicius replies by citing an idea from Aristotle *Physics* 3.6, 206a33-b3; 3.8, 208a20-1. Temporal objects, like the Olympic games, or a day, or the generations of men differ from spatial objects, in that, as one gets more and more members of the collection, the previous members perish. Simplicius asks how many past years would exist in a beginningless universe. Not a more than finite number, but none. Never does more than one year exist at a time. This answer is unsatisfactory, because Aristotle's objection to a more than finite number of anything would apply even to defunct years. Even in a more than finite collection of *defunct years*, parts of the collection would be more than finite, like the whole.

Successive entities

One interesting aspect of Simplicius' discussion, however, is his talk of divisions (*diairêmata* – I take Diels' *diairemata* to be a misprint), 506,11, of temporally extended objects. Surprisingly, it turns out that a division is not superseded by the next division, but rather is said to be prolonged (*auxanesthai*, line 13). This is unexpected, given Aristotle's claim that what you had before ceases to exist and Aristotle's point could well have been put by saying that the Olympic games, unlike an extended substance, has temporal parts. The Middle Ages took up this distinction between successive and permanent entities (see Cecilia Trifogli, *Oxford Physics in the Thirteenth Century*, Leiden 2000, ch. 4). But the distinction was sometimes denied, e.g. by Ockham, who thought we needed to postulate only permanent entities (see Marilyn McCord Adams, *William Ockham*, Notre Dame Indiana 1987), and by David Lewis, who sees even physical substances as successive entities with temporal parts.

Translator's Note

This translation follows the text in *Simplicii in Aristotelis Physicorum Libros Quattuor Priores Commentaria*, edited by H. Diels in *Commentaria in Aristotelem Graeca*, vol. 9, Berlin, 1882, except as indicated in the footnotes and list of textual emendations. The text gives only the beginning and end of the lemmata; I have added the rest of Aristotle's text in square brackets for the convenience of the reader. I use my own translation, partly to ensure uniformity of translation between the lemma and the commentary, partly because I give as the text of Aristotle what Simplicius certainly or probably read in his text, not the received text. I have added a note where these differ. Otherwise I have used the text printed by Ross in his edition to supply the missing text of Aristotle.

I have generally translated *peras* as 'limit', *peperasmenon* as 'limited', *apeiron* as 'unlimited'. The translations 'finite' and 'infinite' are often used, and with reference to mathematics are unexceptionable. But Aristotle is primarily concerned with the question of there being unlimited perceptible natural magnitudes and holds that all magnitudes must be physically limited or bounded by something else; so sometimes I have used the translation 'bounded'. Aristotle even thinks it permissible to speak of roads with a way out as unlimited, or unbounded, those without a way out as limited, or bounded, though he does not himself do so; but it would be absurd to call such roads infinite and finite.

My translation has been saved from many errors and in general much improved both by anonymous critics and by the editorial staff at King's College, London, and I am grateful to them. No doubt the improvement would have been yet greater if I had always followed their advice.

Textual Emendations

395,27	Order of *kinêsis* and *metabolê* as in MSS aF (see n. 9)
399,3	Reading *mêden* instead of *mêdenos*
402,29	Reading *ei estin* for *epei mê estin*
403,21-2	Punctuating *meiôsis. kai allo hê pou katêgoria ... metabolê; hôste* (cf. n. 40)
406,2	Reading *akinêtôn* 'unchanged' (the received text of Plato) for *kinêtôn* 'changed', which is absurd
416,8	Perhaps reading *hadrunsis gêransis* (cf. n. 77)
417,15	Perhaps a lacuna (cf. n. 81)
417,16	Accenting *tinos* and *ti* which Diels leaves unaccented
418,23	Omitting *kinei*
420,19	Reading *sunkhusin* with Giacomini instead of *sunkrisin*
427,17	Omitting *dunamei*
431,14-15	Reading *alla gar horistê ouk esti* with Diels (app. crit.)
432,10	Supplying *loipon* with Diels (app. crit.)
432,13	Reading *ta dunamei kath' auta* for *to dunamei*
436,14	Punctuating *kineitai de kai to kinoun hôsper eirêtai pan, to dunamei on kinêtikon* (see n. 145) for Diels' *kineitai de kai to kinoun, hôsper eirêtai, pan to dunamei on kinêtikon*
441,14	Reading *ousia* for Diels' *ousiâi* (dat. sg.)
451,28	Deleting *diairei ... pseudos* as a marginal gloss with Diels
458,16	Reading *tautais* instead of *toutois*
463,33	Omitting Diels' addition <*arkhên ekhei*>
464,27	Reading *eti arkhê ginoito* (*Phdr.* 245D) for *ex arkhês ginoito*
465,30-1	Reading *tinos* instead of *titos* (a misprint)
467,37	Reading *holên* for Diels' *hulên* (a misprint)
470,11	Possibly delete *ê* and read *tôn gar zôôn tôn enüdrôn*
475,31	Reading *epistêmonos* (with Vitelli) instead of *epistêmês*
482,13	Reading *prolambanei* (with Vitelli) instead of *proslambanei*
486,30	Reading *kata* instead of *para*
490,31	Reading *tode ti* for *to de ti* (Diels)
494,22	Reading *aitian* instead of *aition*
504,23	Reading *to hen men* at 207b2 (with MSS E and I) for *to en men*
505,16-17	Reading *duo kai tria* with Diels (app. crit.; cf. n. 340)
507,33	Reading *aïdiôi* (with iota subscript) for Diels *aïdiô* (a misprint)

508,38	Reading *aporia* with MS F for Diels' *apeiria*
509,21	Reading *hoti* in the lemma; Ross has *dioti* in 207b26
511,15	Reading *ou monon* <*ou*>
512,16	Reading *ti kai exôthen* (Diels' conjecture) for *tis kai exôthen*
515,33-4	Perhaps read *prolêpsis* for *proslêpsis* (cf. n. 368)

Simplicius
On Aristotle
Physics 3

Translation

The Commentary of Simplicius the Philosopher on Book 3 of Aristotle's *Physics*

In the previous book Aristotle discussed causes and said that nature was the efficient cause; he defined it as the principle of change and reasonably next teaches about change.[1] For if nature has its essence in this, in being the principle of change, it is altogether necessary for one who is to understand what nature is to have knowledge about change as well, which is included in the definition of nature.[2] Moreover, if knowledge of related terms is simultaneous, and a principle and that of which it is the principle are related terms, and nature is the principle of change, it is impossible for one who is ignorant about change to understand nature.

So if it is necessary for the natural scientist to know what nature is, and for him who is to know what nature is to have a knowledge of change, the account of change, which is, as I said, included in the definition of nature, necessarily follows immediately after that of nature. On account of the same natural order it is necessary to discuss continuity, since change is an example of continuity and is itself continuous, and continuity is included in the definition of change. For one cannot understand a defined term without understanding each of the terms included in the definition. Again, since in defining continuity we make use of the unlimited and say that the divisible without limit is continuous, it is necessary to discuss the unlimited also if one is to understand change and nature which is the principle of change.

But since everything natural is bodily, and every body and bodily change,[3] the subject of this account, occurs in a place and at a time, the account of place and time is essential for the natural scientist who is to understand change and nature. To some, time seems to be the motion of the heavens,[4] and to others an everlasting moving image,[5] so that in all respects time is akin to change. But since some, such as Democritus, suppose that place is the void and say that change takes place completely in the void,[6] it is necessary also to inquire into the void, whether it does or does not exist at all. Thus if things that change do so in a void, as being in place, then for these reasons the natural scientist must inquire about change, and continuity, and infinity, and place and time, and the void.

However, since these are common to what is natural and universally present in all, the investigation which gives an account of what is common to all things, such as that now lying before us, must also come first as an overview for the other investigations which teach about the attributes of each special case. For every natural body changes and is either limited or unlimited, and it changes its place either as a whole or in its parts, like the fixed sphere, and change is measured by time. So these things are common; but the account of the void is also common, because some think that place is a void and the void is a place deprived of body. So Aristotle handed down these matters even more clearly than did the commentators, beginning at once in Book 3; in it he teaches about change and about the unlimited, in Book 4 about place, about the void and about time. For he has for the present set aside continuity, since even apart from continuity the account of the unlimited is necessary for the natural scientist because natural bodies and natural changes must be either unlimited or limited. However he will later discuss continuity also.[7]

It should be known that at many places there are different readings in the text of this book. But we must move to the discussion of the text passage by passage.

200b12 Since nature is the principle of change (*kinêsis*) and transformation (*metabolê*), [and our project is concerned with nature, it must not remain obscure what change is; for necessarily if that is not understood then nature also will not be understood. But when we have defined change we must try to approach matters that are consequent in the same way. Change seems to be among things continuous, and the unlimited first makes its appearance in the continuous. Therefore those defining continuity will often need to make use of the account of the unlimited since continuity is unlimited divisibility. Furthermore there can be no change without place and void and time. So it is clear for these reasons, and because they are common to all things and universal, that we must embark on an inquiry about each of them. For the study of the specific is subsequent to that of the general. First, as we said, we must inquire about change.][8]

He has set 'change' and 'transformation' side by side here, both because he has not yet distinguished how change differs from transformation and because transformation is more general and change more specific,[9] which he makes clear at the beginning of Book 5.[10] But perhaps he introduced change and transformation in conformity with later distinctions; for he will say that what is substantial is transformation and not change, but what is quantitative, qualitative and of place are all changes. So he said strictly that nature was the principle of change and transformation, since it is the cause of both the substantial and the rest. Alexander notes that 'in saying that change is continuous he places it firmly in the category of quantity; for the continuous and the discrete are species of quantity'. But in the *Categories* he does not place it in quantity, and there a little later he will assign change to relation.[11] 'So', he says, 'it is either for this reason that he said that "change seems to be among things continuous", as not accepting it, or more likely change is in a way both a quantity and continuous, in a way relational, considered in different respects.' Change itself is a quantity, but what is changed stands in a relation, as being in some relationship with what changes it. 'The word "seems" is a sign', he [Alexander] says, 'that he [Aristotle] is beginning from the observable and obvious.'[12]

Having said that the unlimited first makes its appearance in things that are continuous, he shows by the word 'first' that the unlimited appears strictly as such in the continuous, and not through some medium. Those features occur as such which are themselves included in the definition of such things of which they also complete the determination. But those features are also as such which through their association with things of the same sort mark off something special to them from things which are present in common in the whole genus.[13] This primarily holds as such of each of these features, and this is another way in which things are as such. In this way being active holds of fire as such, since being active or passive holds of every natural body. He added 'often' because we do not always define the continuous in that way, by saying that the continuous is the divisible without limit, but there is another definition which he himself set out in the *Categories*, saying that that is continuous of which the parts join at some common boundary.[14] 'But', says Alexander, 'we do not make use of the unlimited when defining discrete quantity.' So in this way also he appropriately said that it 'first' made its appearance in the continuous, or else because the limitlessness of the discrete also has its origin from the continuous. For the cut of the continuous without limit permits increase without limit to the discrete, but still the continuous does not receive its limitlessness from the discrete. For the discrete is not unlimited through being composed of unlimited parts, since there are not actually unlimited segments but its limitlessness comes from unending addition. But unending addition arises from the inexhaustibility of the cut of the continuous. For nor does the continuous contain an actual unlimited, but by being divisible without limit. For nothing is actually unlimited but continues without limit. So why does the discrete receive its limitlessness from the continuous but not the continuous from the discrete? Is it because that which is divided without limit can confer increase without limit on the discrete, but what is increasing without limit by division still does not give its unlimited divisibility to the continuous?

397,1 But when he says 'Furthermore, there can be no change without place and void and time' he does not say this as what seems to be the case to him but because the circle of Democritus accepted the views of the natural scientists before him that motion was through the void and that the void was place deprived of body.[15] Thus the discussion
5 of the void is necessary to that about place. So the discussion of the void is necessary both for those who affirm the void and for those who deny it.

He reasonably first gives an analysis of change, because, after nature, the account of change naturally follows, since it is included in the definition of nature, and because time and place come in through change. For there must be time since what is changed is
10 changed at a time, and place since what is changed is changed somewhere. However it is not necessary for there to be change because there is time or place. The void will be shown to be completely non-existent. Also the unlimited does not exist at one time, but its being is in becoming, and so it too has its existence in connection with change. Further the account of change is worth attention also because
15 it was neglected by earlier philosophers and Plato said little about it. Also all the things that there are are either changes, such as coming to be[16] and alteration, or are origins of change such as god, soul and nature, or things being changed, such as bodies both simple and compound, or the media of their change, such as time and place. But it is worth considering whether it is not necessary, as there is time if there is change, also for there to be change if there is time, since time
20 is the number of change.[17] But there is not change because there is time, but time because there is change. And now he outlines what the essence of change is. What follows on the account of change is a large number of varied theorems which are elaborated in the last four books.

25 **200b26 There is indeed that which exists only in actuality [and that which exists both potentially and in actuality, a particular thing, a quantity, a quality,] and similarly in the other categories of being.**

He is preparing to demonstrate (1) what change is and to draw together its definition so far as possible; (2) in what it is, that it is in what is changed; (3) that 'change' has many senses, and (4) that each
30 species of change is divided into opposed species, e.g. that in existence into coming to be and ceasing to be, that in colour into becoming white and becoming black,[18] and generally that in quality into qualities opposed to each other, that in quantity into growth and diminution,

that in place into rising and falling and so on. In preparing to give this instruction about place, as also about other matters which contribute to the account of its essence and as in the meantime sufficient for a general understanding of change, he first assumes four axioms, of which each contributes in order to each of these statements.

The first of the axioms is arrived at by division as follows: of what exists some do so 'only in actuality, some potentially and in actuality, one being a particular thing, another a quality, another a quantity, and similarly in the other categories of being'. Existing only in actuality holds of the essences and the substantial activities of things ungenerated. Existing potentially and in actuality holds in all the categories of things generated. Further only the immaterial and primary forms are simply in actuality, for they can come to be nothing else beyond what they are from the beginning; but the compounds of form and matter exist both in actuality and potentially – in actuality in so far as they have in actuality some form and condition such as the bronze, potentially because they can receive another also, that of a statue. But the etherial body[19] exists by its essence only in actuality, for they would never change their essence, but they partake in a way in potential being through their change of place and the varying relationships between them that come about because of this change.[20] For they are not everywhere together, nor is the composition of the influences of them all always the same, as the varying results of their varying configurations show.

This division is inescapable; for the complete one would be that some among entities would be in actuality only, some potentially only, some in both potentiality and actuality. But there is nothing which is only potential in existence. For everything that is potential is something in actuality, which has an unfulfilled potentiality within itself, through which it can acquire something beyond itself, as that which is potentially white is in actuality a body which is potentially white, since it is not yet white but is of a nature to be transformed into something white, and is on that account said to be potentially white. And even if in existence such a thing is one, still conceptually the substrate which we say has also the potentiality and nature, e.g. a colour, is one thing, the incomplete potentiality in it another, which is what it is potentially, such as being a visible thing; for the one is a quality, the other a relation. Similarly it is one thing to be snow, another to be the meltable. For, if they are the same, to be snow and meltable would also be the same. For what are the same as the same thing must be the same as each other. So also being divisible and being meltable would be the same thing, so that also being divided and being melted would be the same. For as are the potential, so are the actual.[21]

Plotinus also demonstrates the same thing at the beginning of his book which he entitled *On the Potential and the Actual*. He says: 'However that which is something potentially must be said to be potentially able to be something after what it is, either continuing after making this or providing itself to that which it is potentially, and after perishing must be called that potentially. For bronze is potentially a statue in one way, water is potentially snow and air fire in another.' In general, as he himself says, 'potentially' cannot be used simply. 'For it is not possible to be potentially nothing,[22] as bronze is potentially a statue. For if there was to be nothing from it and nothing additional to it after it, and it could not become something, it would only be what it was. But what it was was already present and not in the future. So what was its potentiality after what it was already? What would it be if only potential?'[23] But having said this at the end of the book, he says that matter exists only potentially, being nothing existent in actuality.[24] But perhaps he wishes the primary existent to be form and that there should be no existence prior to form, wherefore matter that lost forms would be removed from reality, and therefore as non-existent would be only potentially what it is. However even the potential and what is in that way non-existent has not completely lost reality. But, in general, if it is potential in such a way as to be transformed into the actual, as water into air, matter would be destroyed, which is not what he wants. But if it is potential so as to remain while it receives the actual, as bronze receives the form of a statue, it is actually something else of a nature to receive the forms. But if it were nothing else but received what was actual, being potentially and actually would be the same, which is not what he wants. For matter is not potentially form, but rather it remains what it is in actuality and is of a nature to receive forms.[25]

But Porphyry does not accept Alexander's punctuation by which that which exists only in actuality is divided off from that which is both potentially and in actuality.[26] 'For', he says, 'having said that some things are actual, some potential, he returns to the actual and says that the actual is the particular thing, the quantity and so on; but he does not divide the potential into categories; for the potential is not tenfold but single; for the highest matter, which is potentiality itself, is one. Whether he takes the actual, which is the complex, such as a statue, or actuality, which is the simple, such as a form, each is tenfold. Also perhaps he says that also the potential is tenfold. For Aristotle himself, as he continued,[27] spoke as follows: 'the actual and the potential having been distinguished in each kind.'[28] That is what Porphyry said, holding that the actual alone is opposed to the potential alone by Aristotle, not to the actual and the potential. And perhaps it was from this sentence that he acquired this understanding of that which says 'the actual and the potential having been distinguished in each kind.' But if the ancient text is the same as the majority of the copies and runs: 'There is that which is only actual and that which is both actual and potential, the particular thing, the quantity etc.', the addition of 'only' clearly shows that the division was into the solely actual and that which is both actual and potential, as Alexander and Themistius received it.[29] For if he opposed the potential alone to the actual, never including them both together, why did he add 'alone' to 'actual'? And how would the interpretation be grammatical when it has the detached phrases 'the particular thing, the quantity'? But if the sentence runs as it is written in some books 'there is something which is actual, something potential, and the particular thing, the quantity are actual' it can also be punctuated as Porphyry received it.[30]

But why does he divide only the actual into the ten categories, even though the potential is seen in all, as Porphyry himself agreed, for he added Aristotle's words 'both the actual and the potential being divided into each kind'?[31] So perhaps the initial division was into the actual alone and the both actual and potential, and he says that the latter is divided into the ten categories. For neither Where nor When nor Posture is compatible with that which is only actual, which is the immaterial and intelligible form, but the both potential and actual is observed in each genus, as in Quality something is sometimes actually white or warm, when it has already received the form, sometimes potentially, when it is not yet so but can become so.[32] This being so, he reasonably said 'both the actual and the potential being divided into each kind'. For this is why he said also at the beginning of the division that the actual together with the potential could be seen in all the categories, but did not say that what was only actual was divided into the ten categories, and he does not now add it save for the sake of a complete division.

It should be understood that he does not divide what is as a kind of thing into the actual and the both actual and potential, but as 'is' being an equivocal word, and also that being potential is one thing, potentiality another. Also being potential is opposed to being actual, but actuality to potentiality. Further, potentiality is a complete preparedness for existence and an unimpeded readiness for activity, conferring actuality, but what is potential is an incomplete suitability for being what it is said to be potentially, which receives its actuality from something else and does not project it from itself; but what is actual is that which can already be active in the way in which it is named. For he is actually a man who is already active as having a human form. But actuality is that which is opposed to potentiality and is the active change that is projected from potentiality, while that change is also active which is in action, action being making and doing. Also from the potentiality which remains internal and is not observable the word 'potential' was introduced for what is viewed as being only in a suitable condition for existence, being as it were a state of existence itself. From 'actuality' comes the 'actual' in accordance with its existence, which is again viewed as actual. This assumption which divides things into those that are actual and those both actual and potential is of use to him, as I said, towards giving his definition of change, as we shall see as we go on.

200b28 Of relative terms, some are so-called through excess or deficiency, [some through being active and passive, and in general as causing and undergoing change; for the changer changes what is changed] and what is changed is changed by a changer.

He adds this second assumption as being useful to him for showing in what things there is change, that it is in things changed. In it he makes a division of the relational, not dividing all the species of relation but recalling only that of excess and deficiency and that of being active and passive, because he himself believes that change is found in the active and passive, but some others that it is found in excess and deficiency, and these he will attempt to refute as he goes on. For change is also a relation, but Plato placed it in excess and deficiency, saying that the causes of change were an inequality and dissimilarity and otherness between the changer and the changed.[33] But Aristotle did not assign it to this class but to that of doing and suffering and, in general, to that of the changer and the changed. For in many cases the activity of the changer, which is doing, will be suffering for that to which the activity is directed. For example, the activity of him who strikes, which consists in striking, is for him who is struck something suffered – being struck; these things are also relational. Striking is related to being struck, and the doer to the sufferer, and being a changer to being changed. He took the potential to be those things in which change occurs, because their actualization, *qua* changeable, is change, as will be demonstrated.

He has opposed the doer to the sufferer according to common word-usage; but since according to linguistic form the done is opposed to the doer and the suffered to the sufferer, but 'the sufferer' is not so called as bringing about suffering, but in the same way as 'the suffered', for the sake of clarity and greater suitability to the subject-matter he returned to 'the changer' of 'the changed'. In these the active is clearly such, and 'the sufferer' as in ordinary use.[34]

The great and the small show that 'excess and deficiency' is a case of relation.[35] There are also many other species of relation, some of equality, some of similarity, some of recognition, such as sight and the seen and knowledge and the known. But he himself only included

those types of relation which are now useful to him with regard to
change, from which he shows that change does not consist in excess
and deficiency, as some thought, but in the changer and the changed,
or the doer and the sufferer, and in general in things which are being
transformed from potentiality into actuality by that which is actual.
That the changer and the changed are an example of relation he
showed through their correspondence, saying: 'For the changer
changes what is changed and what is changed is changed by a
changer.'

200b32 There is no change in addition to things. [For what is
transformed is always transformed either in substance or in
quantity or in quality or in place, and there is nothing common
to these to be found, as we say, which is neither a particular
thing nor a quantity nor a quality, nor anything in the other
categories.] So there is neither change nor transformation of
anything beyond the above, since there is nothing beyond them.

He adds this third assumption that there is no change beyond things
changed, by which he confirms what was said before, that change is
in things changed or changers. For, if there is no change in addition
to things, change is in things that are changed or that change them,
and there could not be a change which was on its own. There will not
be change in itself, either as the idea of change being the change itself
but the change of nothing, as Plato seems to say in both the *Sophist*
and the *Parmenides*,[36] or as a substance that is change itself, as Plato
defines the soul, calling it a self-changing change.[37] For if change is
in things changed or things that change them change is some attribute and not a substance to be viewed on its own. Also he will show
that it is in things changed.

So he both confirms these points through the lemma stated, and
still more clearly and more specifically that 'change' has more than
one sense. Also there is no common genus beyond the particular, or
generic change itself, since change is only in things themselves,
occurring in four ways – in substance, in quality, in quantity and of
place. Of these there is no common genus, since they themselves are
considered to be primary genera. Also for this reason, as we said, he
recalls that it has been demonstrated that being is not the common
genus of the ten categories. For if there were something common to
those categories in which there is change, change would be primarily
in that common genus and would itself be a genus of the kinds of
change subordinate to it, and change would no longer be one of those
things that are equivocally named. For change would be of the same
status as those things in which it is, if it is outside things.[38]

He demonstrates inductively that change is not additional to things nor to be viewed outside them. 'For', he says, 'what is transformed is transformed in substance or in quantity or in quality or in place', and it is not possible to conceive of anything which is transformed but not in one of these ways. It would be clear to anyone who takes note of the fact that in refuting change as a genus Aristotle proves that change is one of the things that are equivocally named that he is not now alluding to the definition of the soul which says that it is self-changing change nor to the idea of change. Neither the idea of change nor the soul is a genus of change according to what the Peripatetics mean by 'genus', even if genera are constructive of species and even if they have the same sort of existence as they do. Perhaps he noticed that some even then would gladly have counted change among the categories, like Plotinus later.[39] It is because of these people that he says that change is not a genus. But if all change is viewed in the category of relation as consisting in changer and changed, how can changes, being in one genus, not be univocally named but be equivocally named? 'Perhaps', says Alexander, 'nothing prevents some things that are in the same genus being equivocally named the same. Surely the Alexanders that are in the category of substance, both the picture and the man, are still named the same equivocally. Also being equal, which is a relation, is equivocally named as between the continuous and the discrete. Thus change also is in the category of relation since each of them is related to something else. However, changes are equivocally named because the things in which they occur have nothing generically in common, but are different genera. For substance is one thing, in which there is coming and ceasing to be, quality another, in which there is alteration, quantity another, in which there is growth and diminution. And place is another category, in which there is change of place; so that the accounts will be different also of the changes in them.'[40] So says Alexander, who has well observed that being in the same genus does not make things to be univocally named, but being that which is signified by a common name and which is under a common species. For the sake of this Aristotle also added 'the account according to the name' in the *Categories*.[41] Therefore a man is named univocally with another, but change in substance is not univocally named with that in quality. I think it is worth observing that, if changes are equivocally named for this reason, that the things in which they occur have no common genus, it is clear that where the same thing occurs in things classed under one genus, even if it be from another genus, nothing prevents them being univocally named on that account. So what prevents equality viewed in the continuous and in the discrete, which are both in the category of quantity, being univocally named? And how, then, is change equivocally named because it occurs in many genera? If it is, relation also, since it is seen in substance and quantity and quality, should be equivocally named and not a genus. 35

Having said that one can find nothing common to the different sorts of change, he made clear what sort of common element he is 404,1 denying by shifting the discussion to the categories in which there is change when he said 'which is neither a particular thing nor a quantity nor a quality'. Genus and species are that sort of common element. For it is not because they encompass everything falling under them that these are said to be common elements, but because they are no particular thing yet make their appearance in all particu- 5 lar things; for example, a skill is not a genus of particulars because it is theoretical, practical and productive, but because it is no particular thing but makes its appearance in all the particulars. But if change is equivocally named, how does he define it? For there are no definitions of the equivocally named, or else the definition of the equivocally named is also equivocal. For since the primary (*arkhê*) is 10 equivocally named the definition that says that the primary is the first in each thing is also equivocal.[42] Moreover the actuality of the changed *qua* changed will be equivocal. For the product of equivocals is equivocal. Therefore he had to add this lemma lest someone who heard the definition of change should think that the definition is given of it as a genus, and so that it should be clear that the account is given 15 through an equivocal expression as being of an equivocal expression.

But since the Lycian philosopher, [Proclus], says that this opinion about change is the one and only disagreement between Aristotle and Plato, the former saying that there is no change beyond things and refuting the view that change is a genus, the latter that change is a single genus of being as are existence and identity and otherness, it 20 would be more seemly to demonstrate agreement in the apparent disagreement if at all possible.[43] We can say at once that Plato, since he treats what are in his works called genera of being in the intellective universe as being the first distinguished causes of existents, reasonably also exhibits change and rest, identity and otherness, and the rest as self-existent. For as in them there is an ineffable unity, so 25 there is also an unconfounded purity.[44] But Aristotle is enquiring into natural and material change and reasonably views this sort of change as being in changing things. For *here* there is no change on its own. That is why he also adds 'for the transformed is transformed either in substance or in quality or in quantity or in place'. That transformations and the categories are strictly among generated things is 30 clearly proved to those who have thought about them. It is also clear from these cases, since he assigns change to things having potentiality.

As of more immediate relevance and more concerned with the matters before us, I think that it is possible to say that Aristotle observed the special nature of change quite remarkably in this respect also. For having assigned to it activity and passivity he said that there was no change apart from things, i.e. apart from the kinds of being. That is why he added 'for what is transformed is always transformed either in substance or in quantity etc.' or how can one observe activity or passivity separately, without something acting or suffering? Plato also knows of such a difference in things, by which some things exist separately while others are observed with other things and in other things.[45]

Thirdly, let it be said that the matter in hand is to show that change is not a genus, just as he also does not believe that being is a genus, since it does not apply to all things in the same way. On this point the friends of Plato agree, since they themselves say that being comes to all things from one, as did Aristotle.[46] Therefore this is one signification of 'genus' – that which is divided into species which share equally in the genus, which Aristotle rejects for change, because the many sorts of change differ in degree of change from each other; but there is another signification of 'genus' by which Plato calls genera of being those things which pervade all things in turn even if they are not present equally nor in the same particular way to all in turn. So it is nothing surprising if Plato calls change a genus according to this signification of 'genus'. But Aristotle, who denies that it is a genus in the other signification, claims that there is no change as a genus. However, he does not hold that even the genera which he accepts exist in themselves or subsist apart from their species, nor the species apart from individuals, when he views them as species and genera or, in other words, as elements in individuals and not as transcendent causes. For in the latter case he holds them to precede those things resulting from them and sharing the same special character.

Fourthly, I think that philosophers should take notice that it is one sort of change that Plato treats as a single genus of being, viewing it as having one signification, and that it is another sort, which is conceptually different, about which Aristotle is now teaching. For the former kind of change signifies the first separation from being into powers and activities that are both vital and intellective which is utterly without transformation, as what is said in the *Sophist* makes clear,[47] from which change seems to be discovered:

STRANGER. What then, by Zeus? Shall we be persuaded easily that truly change and life and soul and thought are not present in the wholly real, that it neither lives nor thinks, but august and holy, without intelligence, it remains without change?
THEAETETUS. We should indeed, Stranger, accept a strange story.
S. But shall we say that it has intelligence, but not life?
T. How could we?

S. But do we say that both of these are within it, but that it does 35
not have them in the soul?

T. How else could it have them?

S. But then that it has intelligence and life and soul, but while animate remains altogether unchanging?

T. That all seems laughable to me. 406,1

S. So, then, do we accept that there is both the changed and change? So the result, Theaetetus, is that if things are unchanged[48] there is intelligence in nothing about anything anywhere.

Observe that this change, which Plato discovered and counts with the other genera, was viewed as existent, because it was shown to exist in accordance with the life and intelligence of the wholly existent. 5
But, since THERE[49] everything, as I said before, has an indestructible and self-sufficient distinctiveness as well as unconfused unity, for that reason change also came to be viewed as on its own and to be classed as a genus of existents. But the change transmitted to our world is ever-flowing transformation, and the potential activity of 10
what remains potential. So it is measured by time, as the former is by eternity; and it has descended among things changeable and lacks its self-sufficient existence.[50] The cause of this difference of viewpoint is, I think, that Plato thinks it right that the paradigmatic causes of things HERE should be called by the same names, but Aristotle avoided such an equivocation, that occurs because we form within us 15
also a similar concept corresponding to the word.[51] But enough about that. We must go on to what comes next.

> **201a3** Each of these is present in all things in two ways, [e.g. the particular thing, since one is its form, the other privation; and in quality, since one is white the other black; and in quantity, since one is complete the other incomplete. Similarly in travel one is up, the other down, i.e. one is light, the other heavy], so that there are as many forms of change and transformation as of being.

He adds a fourth assumption, that each of the things mentioned in which there is change, i.e. substance, quality, quantity and place, is 20
present in everything that has them in two ways. For what partakes of substance does so either as having form or as deprived of form, and what partakes of quality partakes of each in respect of one of the opposites in the category of quality; for in respect of colour it partakes of white or black, in respect of flavours of sweet or sharp, in respect of tangible qualities of warmth or cold. The same account holds in 25
respect of quantity and place, since clearly everything in between has

the role of an opposite to opposite. For grey is white in relation to black and black in relation to white.

The commentators accept that form and privation are not mentioned with respect to substance alone, but also with respect to quantity, quality and place, since in all of them the inferior of the opposites seems to be a privation. 'For as in the case of substance the statue is form, privation the shapelessness of the bronze, so in the case of quality white is form and black privation, excellence is form and badness privation, and in the case of quantity there is the complete and the incomplete, and in the case of place what is above and light is form, what is below and heavy privation. Hence the inferior have a lower and heavy nature, old age bends us down, disease weighs us down, and the divine are above, the mortal below. For excellence (*aretê*) is so called from raising up (*airein anô*) and badness (*kakia*) from bending down (*klan*) but bending (*klasis*) and being bent are inferior.' Such is the philology of Porphyry.[52] But perhaps Aristotle assigned 'form' and 'privation' more strictly to substance, and to the other categories assigned subordinately 'opposites'. For he contrasted the incomplete either with the complete as being lesser or with the moderate as greater, saying both in respect of quantity, and taking suitable examples in the cases of quality and place.[53] But if someone were to say that in each of the oppositions the superior was more akin to form, the inferior to privation, he would do so not implausibly; for it is obvious to all that form and shape are in the highest degree suited to substance. So since the shape is something of two sorts in each kind of thing in which change occurs, so in each of these there will be two sorts of change and transformation from each of the opposites to its opposite. For in substantial change one sort is coming to be, the other ceasing to be; of change in quantity one is increase, the other decrease; in change of quality it is becoming white and becoming black, and in the case of the other transformations in quality the changes are named from the qualities; in change of place the one is rising up, the other descent below. In every case the journey to the superior of the opposites, which has the status of a form, will make it itself more analogous to form, the journey to the inferior and privative to privation.[54]

'Having shown', says Alexander, 'that each of the things in which there is change is present in all things in two ways', he reasonably adds 'so that there are as many forms of change and transformation as of being.' For, if there is no change in addition to things, change would reasonably be divided correspondingly to the things in which it is. Still more, if change from either thing in each kind to the other is a form of activity, it is clear that it itself will be double. 'But', Alexander says, 'this phrase would be more appropriately placed before "each is present in two ways", being joined up with "there is no change in addition to things".' But perhaps, if he had now to add the double form of changes, he would first well show that the double form was in the kinds of being in which there was change and through that would double the kinds and then would add 'so there are as many forms of change and transformation as of being', i.e. that these also were doubled. For since form and privation are in substance, so changes in substance are also double – that to form being coming to be, that to ceasing to be being privation, and similarly in the other categories. But if, as Alexander claimed, this phrase ought to have been placed before that saying 'each is present in two ways', what further need would there have been to add it?

But he now speaks of species of change and transformation not as distinguished within one genus but as different particularities. He well said not just 'change' but also 'transformation' because he will later inquire about that of substance which is coming to be and ceasing to be and perhaps this is not change but is transformation. So if anyone is puzzled how he said that there are as many species of change and transformation as of being, given that he will show that change is in only four categories, those of substance,[55] quality, quantity and place, let him know that after saying 'each of the genera or species in all of which change is doubly present according to the opposition in each', since there are as many kinds of change as of things in which there is change, he added 'so there are as many species of change and transformation as of being'. He does not include all kinds of being, but that which has a corresponding change. That is being in which there is change. This is that which is included in the four kinds of being.

But it is worth inquiring why, since potentiality and actuality are in all the kinds of being and since change is the activity of the potential as such, change is in only four kinds of being, substance, quality, quantity and place, but not also in the others although the actualization of the potential which remains potential is in them all. This distinction was made in the *Categories* and he will take trouble to demonstrate it in the fifth book of this work.[56] For the transformation of that which is on the right but is of a nature to become on the left is the actualization of the potential that remains potential until when it is transformed. And he who is seated but is of a nature to lie down as well and who is being transformed from one to the other, the potentiality remaining in the transformation, falls under the definition of change through such a transformation.[57] We shall find in the case of the other categories also an abundance of transformations to the different species falling under the genus to subsume under the definition of change. But Aristotle in the fifth book of this work states the reasons why change occurs in only the four or the three categories,[58] and Aristotle's commentators postpone to that place the solution of this problem. But perhaps it is just as well to examine now briefly the explanations given there for the sake of those who do not readily postpone the solution of problems to another occasion.

So Aristotle, having judged that changes and transformations come about from opposites to opposites, or from intermediates which have the status of opposites, consequently says: 'In the case of substance there is no change since nothing exists that is the opposite of substance; nor, however is there in the relative, for it is possible for it to be true that when one of the related is transformed the other is in no way transformed.'[59] For example, he who was formerly standing on the right will be himself on the left when the one on the left moves over to the right, without himself being transformed in any way. 'So such a change is incidental. There is neither change in agent and patient nor in anything that is changed and that causes change, because there is no change of change nor generally transformation of transformation.'[60] So having given the accounts of these categories in that way Aristotle omitted the categories of posture, state and time.

On the one hand, he himself clearly admits in the case of substance that, even if there is no change because there is no opposition, still there is at any rate transformation, since he calls coming to be and ceasing to be species of transformation. But in the case of relatives why would there not be change since there is opposition, as Aristotle himself agrees in the *Categories*,[61] where he says 'Opposition occurs in relatives; e.g. excellence is opposite to badness, each being relative, and so is knowledge to ignorance'? For even if when the one on the right moves the one formerly on the left comes to be on the right without moving and thus is transformed incidentally, still first he who formerly stood on the right, and moved not incidentally but himself, exchanged his position on the right together with his place, having come to be on the left. Also, since relative position derives not from one of them but from both, he who did not move gained the relative position not incidentally but as such, as it was changed together with him who changed his place. It is just as if someone by pushing down one end of a rod makes the other move upwards, not incidentally but as such, because there is one motion of both the ends.

Alexander approves of this view, as one hears from what he says a little later, as he explains the passage which says 'therefore change is the actualization of the changeable *qua* changeable'.[62] For he says: 'Or, even in the case of the relative, on the one hand what is potentially double has not been changed, but what was potentially half was changed, in order that the former should become double. On the other hand, the potentiality of the potentially double was not another potentiality from that of the potentially half, so that with regard to the actualization of the potentiality the change in them also produced the different relationships of the related, since there was a single potentiality with regard to the subject matter.'

But, if acting and being acted on are changes and there is consequently no change in those categories, since there could not be change of change without an infinite progression, then, first, if acting and being acted on admit of opposition, as is stated in the *Categories* (11b1) where he says: 'For heating is the opposite of cooling, and being heated of being cooled, and being pleased of being distressed',[63] then it is also possible for there to be transformation from one of the opposites to the other. For one who enjoys food might when glutted and replete be distressed, and he who is being warmed may be made cold and air that was formerly warming may be transformed to chill. So if this is the case why should there not be transformation of the change in the case of acting and being acted on without an unlimited progression, since the transformations are from forms to forms? For we are neither warmed nor chilled without limit. Again, if walking were to come to be and cease to be, how would it not be a transformation of transformation? For walking is also a sort of transformation, and the beginning and cessation of walking are themselves also a sort of transformation. Or, if this is so, since the transformations themselves also which did not previously exist come to be and cease to be at a time, will there be transformations of them also, and so without limit? And if that is absurd, perhaps the absurdity is not removed by there not being transformation of transformation. For that is still more absurd, since obviously I might exhibit building and walking and growing warm, when yesterday, perhaps, I did not but do today, which cannot be so without transformation. For if I am being transformed with regard to them, still they also, being certain forms and, as it were, the substrate, are transformed from not-being into being and from being into not-being. But these changes also should not be taken as the substrate. But, alternatively, this is not simply change of change as such. For the agent is not changed in so far as it acts, while what is acted on, if it were to change from being affected in one way into another, still it is transformed as a certain sort of change, not as being simply a change. So if this account has merit there is both opposition and change or transformation from one thing to another in the category of acting and being acted on. There is not, however, in this case transformation in their transformation, since such transformations are not forms as substrates but as merely transformations: nor are they a change of change, but of a certain sort of change. Further distinctions need to be made about this.

But why should there not be change or transformation in the category of time? For as I am transformed from being white to black and from place to place, so am I from yesterday to today and to tomorrow. How does this differ from changing from being in the

Lyceum to being in the Academy? For even if the parts of time cease to be, that is all that does so. For, even if the Lyceum were to cease to be when I walked to the Academy, the change would not cease to be of place. And the white also ceases to be when I am transformed from white to black. But if someone were to say that change of place is a kind of travel through place and a journey, but nothing travels through time, but rather time passes by together with things as they travel; for old age and the like are all alterations and affections of bodies in and concerning themselves, but time passes together with them in a different sense. If, then, anyone were to say that, he seems to me to rely very much on perception and therefore not to notice the activity of time; while he admits change of place, since place is perceptible, he denies transformation in time since time is not visible. However one should bear in mind that time is continually acting on us, and, as in the case of transformation from white to black, so transformation from yesterday's experience to today's is not *qua* from experience to experience, as from youth to age, which is through the altered condition of the body, but is through the temporal reference of yesterday and today. Eudemus, the closest of Aristotle's associates, also bears witness to this account in the second book of his *Physics*, where he says, after stating the other four types of change: 'For time also accompanies things that potentially change in the way that they can; for everything changes in time.'[64] Unless, of course, someone should say that change was in time but not in respect to time. Some say that the fact that there is no rest in respect to time is a sign that there is no change specifically in respect to time; for nor is any time the same as another, but that which approaches is always another. So nothing will remain in the same time, but some things remain in the same place and with the same quality. But one must consider whether this is a sign of the opposite. For, if no time is the same as any other, but that which approaches is always another, it is clear that time has its being in coming to be and ceasing to be. How could that in yesterday's time, when it comes to be in today's time, not be said to be transformed in respect to time, even if yesterday's time has ceased to exist?

But also with regard to posture, why should it not be said that what is transformed from being supine to being prone changes position, since being prone and supine are different postures and these are contrary to each other? For the transformation it undergoes in relation to the surrounding place is one, that in position, and the shape of its position is another. Similarly why should not he who is transformed from being armed to being unarmed[65] be said to be transformed in state?

Also why should not the substrate be said to change as well in those categories in which there is opposition? In those in which there is not, privations serve best as oppositions with regard to the substrate, as being weaponless is contrary to being armed. And then, even if there be no antithesis, it being the case or not with regard to the category would suffice. For there is also transformation in substance, although there is no opposition. For the opposition which is common to all the genera, which is that of form and privation, is sufficient. Thus he who was formerly armed, if he becomes weaponless, is transformed from possessing to not possessing arms, while it is possible also to be transformed from being more heavily to being more lightly armed.

It is in general worth noticing that all the categories, viewed in coming to be and existing in particular things undergo transformation. For as substance that is HERE is generated and perishing, so are qualities and quantities and the other categories.[66] So, as even if substance could not be said to be changed since the transition is not from substrate into substrate (for what is changed needs always to be a certain substrate), nonetheless substance is admitted to be transformed as such and not merely incidentally, since it comes to be and ceases to be, so also each of the other categories would be admitted to be transformed as such, since it comes to be and ceases to be, as for example being white, and three feet and on the right. And the substance which possesses these would be strictly said to be changed in respect of them, since the object[67] comes and ceases to be in these respects, while the substrate continues, which Aristotle says is a property of things that are changed. But if there is also potentiality in all the categories, while change exhibits the actualization of the potential as of that sort, why should there not be change in all of them? Generally it is clear to all that he who comes to be on the left from being on the right is transformed, as also a friend from being a foe, and also he who turns from heating to cooling and he who turns from being heated to being cooled, does so not only in respect of quality but also in respect of the different and opposite affections experiences; also he who goes on from yesterday to today and he who shifts his position from being prone to being supine and he who becomes weaponless from being armed. Since each transformation is with regard to some sort of being, and these are divided together with the categories, let someone say in respect to which other category each of these transformations occurs, if not its own. I know that it seems rash to speak out contrarily to Aristotle, but, until we are able to recognize accurately the explanation of this sort of classification which he makes, we are sufficiently encouraged by the agreement of Eudemus quoted with regard to time,[68] and still more by that of Theophrastus, who clearly views change and transformation as being in all the categories. At any rate he says in the second book of his *On Change* that 'it is more appropriate, as we say and as is the case, that the actuality of the potentially changeable *qua* changeable should reside in each of the other categories, such as substance, quality, quantity, travel. For alteration, growth, travel, coming to be and their opposites are so'.[69] But in the third book he wrote as follows, still more clearly, as I think: 'In the delimitation of change we say that there are as many species of it as there are categories. For we say that change is the actualization of what is potential as such.' In the same book he also says this: 'There is not change of the relative in its account, but there is of that which is potential. For its actualization is change and as such.'[70] But these matters must be inquired into further as we follow on the trail of Aristotle's views.

201a9 Now that the actual and the potential have been distinguished in each kind, I say that change is the actualization of the potential as such.

Having made assumptions useful for the study of change, he has at once used the first that he made, or rather a portion of it, to contribute to the definition of change. For the distinction was that there is both that which is only actual and that which is potential and actual, which was said to hold in each of the categories. So he says that now, after the study of this that is both potential and actual and its division into the potential and the actual, 'I call the actualization of the potential, as being such, change.'[71] For in the category of substance there is the potentially man, e.g. the seed and the actually so, e.g. Socrates, in quantity being two feet long is potential and actual and in quality being white is potential and actual. And so in the cases of the other genera. That he marvellously defined change we may learn from this: for what is actually that which it is said to be, while it is in that state, would not be said to be changing in that respect; thus a man, so long as he is a man, would not change in regard to manhood, nor, if he is actually white, does he change with regard to whiteness so long as he is white. But if a man who is actually white were to be potentially black,[72] since he was capable of becoming black, when the turning from whiteness to blackness occurred in him, as was natural, i.e. in accordance with his capacity to become black, he would then be said to turn black. And again, when the blackness has come about it then remains constant in him and he is no longer changing with regard to blackness, but is actually black. Thus nothing is changed *qua* actual; nor indeed *qua* potential: when it remains potential and merely suitable it would not be said to be changing. But when it is being transformed from being potential into actuality while retaining its potentiality, then it is said to be changing. So he reasonably added 'as such' in order that to emphasize the actualization of what remains potential. For when there ceases to be potentiality there is no longer change. For what has come to be in actuality is in static rest in so far as actual, but not in change. Nor, however, if something is still only potential is it already changing. For what is buildable, which is that which can be built, is unchanging while it remains inactive with regard to being built, but when it is active with regard to being built, but still retaining its capacity to be built, then it is being changed, i.e. when it is being built; and building, which is the actualization of what is being built, is then a change. For until it has been built it is being built, and it is active and changing in respect of its potentiality.

It is worthy of note that Aristotle, when defining change at the beginning, said it was the activity (*energeia*) of the changeable as such, but Alexander, Porphyry, Themistius and the rest, when commenting on the definition, read Aristotle as a little later calling it the 'actualization' (*entelekheia*).[73] For in some copies they found the text to read 'change is the actualization of the potential as such'. So they take activity to be equivalent to actualization in the definition of change, as being the same in Aristotle. But perhaps Aristotle understands actualization with respect to completion (*teleiotês*), and if perhaps he applies it to an *energeia* it is not to any chance one, but a complete one, because each is then taken in its completeness when he states its own complete natural activities. That is why he defined the soul as the actualization of a natural organic body potentially having life, not because the soul was an activity, but because the body's completion was through it. So it was not pointlessly that he immediately called the change of something that was incomplete an activity and not an actualization. For Alexander well pointed out that if it was also of the potential that change was said by Aristotle to be the actualization, it is so called in so far as its activity is a completion of its potentiality, as in the case of dispositions the completion of the disposition is activity in accordance with it. But in the case of a disposition the activity does not destroy the disposition, but makes it more complete, whereas the activity of the potential *qua* potential that makes complete its actuality also destroys the potentiality. So there could not strictly be actualization of the potential; for, as I said, that word shows continuity to completion, and, when it strictly said of an activity, it is not exhibited with regard to any chance one but of the one that is complete, which is static in actuality, and exists in actualization. So actualization in the strict sense is of two types, one as static complete form, in the way that the soul is said to be an actualization, the other as activity (*energeia*) in respect to that form. But if sometimes activity is called actualization without restriction, this also is so insofar as it exhibits each thing as active according to its own nature, whether the nature is complete or incomplete. But Porphyry says that change is an incomplete actualization but a complete activity.[74] However, if it is an activity of the potential and the potential is incomplete, how could that actualization of the incomplete be complete actualization? For in general, if there is incomplete actualization, it will be of the incomplete. Aristotle himself, too, says a little later[75] 'and change seems to be a sort of activity, but incomplete'. It should be noticed that Aristotle has applied the word 'activity' in common to both the agent and the affected. For it is in common that in the case of both the transition from the potential to the actual is change. And even if he selected what is built as an example, still it is truly said of the builder, since the actualization of the potential is change. For the builder when inactive is potential, but when active he then is in change with regard to house-building. Also building is change, since it is activity of the builder and an affection of the house. Aristotle called them both activity with a common name, in so far as the potential is brought to actuality through both. But perhaps the agent also, when it is transformed from potentiality to actuality, is changed as something affected and not as agent, and for that reason Aristotle inserted all his examples as of the affected. For the activity of the agent *qua* agent is complete.

201a11 For example the actualization of the alterable, *qua* alterable, is alteration, [that of what can be increased and what diminished (for there is no name common to both) is growth and diminution, that of the generable and perishable is coming and ceasing to be,] that of the moveable is travel.

It has already been said that change is not predicated of the many kinds of change univocally but is among the terms with many senses. Since it is such, its definition must also be taken equivocally. So, to make things clear, he presented each kind of change, saying e.g. 'of the alterable, *qua* alterable, alteration' and so on. Since names are lacking in many cases Aristotle well exhibits that in the case of quality there is a common name assigned, which is 'alteration'; for whether it is from white to black, from black to white, from warm to cold, from cold to warm, and in the case of all the other qualities, a common name of the transformation is assigned, which is 'alteration'. Similarly in the case of transformation of place the common name of 'travel' is assigned. However, in the cases of quantity and substance a common name is no longer assigned to the transformations in each direction, but in the case of quantity we call the increase from less to greater growth, and the opposite diminution (*meiôsis*), or, as he calls it, decay (*phthisis*), not using the term to mean the same as ceasing to be which leads to non-existence, but indicating diminution; but in the case of substance that from non-existence to existence is called coming to be simply,[76] the opposite ceasing to be.

201a15 That this is what change is [is clear from this. When the buildable, *qua* being so called, is being actualized, the house is being built, and this is house-building. Similarly in the case of learning, curing, leaping, aging] and ripening.[77]

Having fitted the definition of change to the equivocally named types of change, he next shows that the definition was well-chosen by a syllogism which has obvious premises that need only clarification through examples. The syllogism is as follows: the actualization of the buildable *qua* buildable and its actuality is building; but building is a change; therefore the actualization of the buildable *qua* buildable is change. For while also what is already built is said to be buildable, still the actualization of that and its actuality is its complete form and actuality with respect to it. And that is no longer strictly buildable, but only by back-reference.[78] That is strictly said to be buildable which still exists potentially and for so long as it does so potentially. For when this is being actualized through its potentiality, which is to be buildable in actuality (after all we say that to be actualized in respect of something is to be just that in actuality), then this actualization, which is building, is change, and similarly in the cases of other things that are potential. So therefore it is true universally to say that the actualization of the potential *qua* potential is change. But it seems rather harsh to call the potential actual when it is being actualized according to its potentiality. But since generally what is being actualized is present, and what is being actualized is said to be in actuality, for that reason he also called the potential actual.

But Alexander says: 'On the basis of what has been said someone might substitute that change is the first transformation of the potentially such, or its first actualization. For the last one is the transformation into completion, through which that which was potential has become actual and finally rests in that respect. The first transformation is the road to the latter.' This, which Alexander said in these very words, is to be accepted, unless one were to say that change was the first transformation of the incomplete potentiality itself, a later one being that to the form, but of that which was potentially and naturally so. For this is transformed through change from being incomplete to being complete, and it exhibits through that transformation an incomplete actualization, so long as it remains incomplete. For change was not said to be of the potentiality or the suitability, but of the suitable as being of that sort.[79]

Nonetheless, we must inquire how he says that change is the first transformation when the discussion is about simple change. For if change and transformation are the same thing change would be explained by itself, just as if someone seeking the definition of change were to explain it as the first change. But if transformation is more universal than change, and has the role of a genus, because coming to be and ceasing to be are transformations but not changes, then firstly that distinction has not yet been made; rather here he includes coming and ceasing to be in his view, but they will be distinguished in Book 5.[80] Then change as now under examination is that which is common and includes both transformation and change. For nature is the principle of that universal change, and he is inquiring into that of which nature is the principle. So if transformation is a part or species of such change the whole will be explained by the part, just as if someone seeking the definition of animal were to say it was man. So in general he who says that change is the first transformation of the potential, transformation to completion the last,[81] since everything transformed is transformed from something to something, from what to what is its first transformation?[82] For it is not from the incomplete to the complete. For that would be the last. But if it is that from not being changed to being changed, there will be change before simple change, which is absurd. So is the first transformation change, but the last, which is to completeness, not a change though it is itself a transformation? However, all transformation is change of the kind which is now under investigation, and the transformation from incompleteness to completeness is through change.

It is worth observing that the examples offered of the types of change are learning and curing as an alteration, the one concerned with the soul, the other with the body; rolling and leaping are travel, one internally the other externally caused; maturing in quantity and growing old in substance are growth and decay, the opposites of which he also gave for consideration. But building can be of substance, maturing and growing old are of quantity.[83]

201a19 Since in some case the same thing can be both potential and actual [but not at the same time nor in the same respect, but as what is hot in actuality but potentially cold, they will act on and be acted on by each other in many ways.] For everything will be at once active and acted on.

Having said that change is the actualization of the potential as such, he well adds what follows to the account. For since natural things are both potential and actual, even if not in the same respect but in one respect or another, perhaps potentially cold but actually hot, or, if they are in the same respect both potentially and actually, as in both respects hot, still not at the same time but now potentially, later actually, or the reverse – since natural things are like that, for that reason they are neither only changed, as is matter, nor only initiate change like the incorporeal, but are changed in so far as potential and initiate change in so far as actual. So, as I said, this passage fulfils also the need to give the explanation why natural things are not only changed but also initiate change. Also it well resolves a problem which emerges about the definition of change. The problem raises a question of this sort in the first figure:[84] some things that are changed also themselves initiate change; but all things that initiate change are actual, not potential; therefore some things that are changed are actual and not potential; so that not every change is the actualization of the potential as such. So this problem is solved by the fact that everything that is changed is changed and acted on in so far as potential, but initiates change and acts insofar as actual. For its potentiality is one thing, its actuality another. So its activity is also other in each respect, since not only acting but also being acted on is called an activity. Yet since he holds that change is in the changed, he called change the actualization of the changed as such. But that the potential is acted on in so far as it is potential and being changed but the actual acts and initiates change in so far as actual is clear from the fact that the activity of each is not in the same respect, but the activity of the potential is not in respect to what it is but of what it is of a nature to be. For the cold becomes hot because it is not hot but cold, and of a nature to become hot. But the hot heats because it is active in respect of what it is, and not because it is not yet so but of a nature to be actualized, like what is acted on. For it is because of what it is of a nature to become that it becomes cold when acted on, but does not heat. So all things that are changed while also initiating change are changed because of their potentiality to be so, but initiate change because of what they are actually. For the cold becomes hot *qua* potential, but reciprocally cools as actual. Also, if it were not potential but only actual it would not be changed nor transformed into something else, but would remain in the state in which it was. For it would not be of a nature to become something else, it would only be active and act as it was, but would not be acted on. But if it were only potentially something and was not anything in actuality it would not be active nor act on anything nor change it; for it would not have anything in virtue of which it would act; rather it would not exist at all. For what each thing is it is so actually. That is why some think that matter does not exist at all, since it seems only to be acted on and neither to act or be active.[85] For what is not active is not actually anything through which it would have to be active. So, as I said, the passage before us fulfils these functions as well.

But perhaps he wishes to add something to the definition of change. For, since change is stated to be the actualization of the potential, lest anyone should think that such a thing is only potential, he will soon add that when that which is actually one thing and potentially another is either active itself or activated by something else in respect of its potentiality such an activity is change.[86] So he reasonably assumes that natural things, in which there is change, are both potential and actual and act and are acted on. But he well said that some things were the same both in actuality and potentiality; for not all things are such. For the intellective and divine things contain no potentiality, because they are actual in essence, as he says himself.[87] However the heavenly bodies even though they are bodies do not have any potentiality in their essence, being exempt from coming and ceasing to be, but they too have a potentiality with regard to change of place; this is not because they are transformed from rest to motion but because all their parts are not everywhere at once but at different places at different times. But perhaps they are also subject to alteration and are affected through some potentiality within them because of their different relationships to and conjunctions with each other. For they assume one disposition through co-presence and another from their triangular or hexagonal or rectangular position, and another from a conjunction which has no definite form, as their different effects through such differences show. But if these, that are themselves changed by the primary causes of change, change sublunary affairs, they are not reciprocally changed by them, and they differ in this way from the sublunary.[88] For these are reciprocally changed by what they change as they initiate change because they are composed of the same elements and made of the same material which is equally capable of assuming contrary predicates. Therefore each is affected according to its matter through which it has its potentiality and works according to its form, through which it is what it is in actuality. For things that warm are reciprocally cooled and those that cool are reciprocally warmed. But perhaps the heavenly bodies are similarly affected by acting on each other and being affected by each other.[89] However Alexander says that the heavenly bodies are called changeable more colloquially, because that is strictly changeable which is able to be changed and not changed. So when he says 'so that the natural changer' he is not speaking of everything natural; for the heavenly body is also natural. But it is neither reciprocally changed by the changed nor is it changeable. But perhaps this also holds when both things are composed of the same elements. For as reciprocally affected and changed, as affected by one and another affection and one and another change, it clearly is also changeable. But this is not as being able both to be changed and not to be changed but as having a nature to receive certain features by which it is not yet affected. For the moon when travelling with the sun is not yet affected by its opposition, though it will be affected and changed with respect to that affection, so that it is also changeable in that respect.

201a25 So some believe that everything that initiates change
undergoes it, [but the facts about this will however be clear from
other considerations.] For there is something which initiates
change but is unchanged.

420,1

He has said that some things are both potential and actual, and
therefore are changed with regard to their potentiality and initiate
change in respect of their actuality. However, not all existents are
like this, but there are things which are only actual and not potential 5
and which obviously only initiate change and are not changed. So he
reasonably added that some believe that there is nothing which
initiates change but is unchangeable, but that everything that initiates change undergoes it, the contrary to which he will himself prove
at the end of this work, to where he postponed inquiry on this topic.[90]
It is clear that those of ancient naturalists who supposed the basic 10
principle to be bodily, whether one or many in kind, were of this
opinion, as also the Stoics among the moderns;[91] for those of old said
that things came about through the separation or composition or
alteration of the principles.[92] But Alexander says that this is Plato's
opinion. Now it is clear from what is said in the first hypothesis of the
Parmenides that Plato says that the one and the good and generally 15
the first cause, which draws all things towards itself as the object of
desire, is unchangeable. For after abolishing each kind of change he
concludes from this that 'the one is unchangeable with respect to
every kind of change'.[93] (139A) Alexander himself agrees that Plato
says that the forms are unchangeable; this is why he believes that he
proves that there is some muddle[94] in the arguments of the Platonists.
'For', he says, 'if intellection is being changed according to the Platonists, 20
to be intellected is to initiate change. So if the forms are
intellected they initiate change; so if everything that initiates change
is itself changed the forms also are changed. But they suppose them
to be unchangeable.'[95] But it is quite obvious to everyone that intellection is not said to be an instance of being changed with respect to
any form of natural change, but in respect of its activity, to which
Aristotle also bears witness for intellect, which he says to be activity 25
in its essence. And also in the *Timaeus* Plato says (28A) that everything that has being always remains the same and uniform; but
nothing that is changed whether in substance or in quantity or in
quality or in place can remain the same and uniform.[96] So that when
he says in the *Laws* (898A), 'Both changing in the same way and
uniformly in the same spot round the same things and in relation to 30
the same things, and according to one rule and order – both reason
and the motion that travels in one place that are likened to the travel
of a rounded sphere',[97] he is not bearing witness to a spatial motion
of intellect, either as a whole or on its parts, but to one of activity,

called change entirely through its vital rising up from being to activity. So Aristotle holds that intellect is unchangeable with respect to natural changes having learnt it from Plato.

However, since the discussion is turning towards Plato, it seems to be turning its gaze to his doctrine about the soul. For Plato, like Aristotle, believes that bodies are externally changed, and says that they are immediately changed by a self-changing soul.[98] Aristotle also agrees that all the heavenly bodies also are immediately moved by soul; for he clearly says that they are animate in the second book of *On the Heavens* where he has raised some fundamental problems about the stars and is about to solve them. He says: (292a18) 'We think about them as bodies and as having the status of units that are altogether inanimate; but we should envisage them as having a share in action and life.'[99] He says that they move with the whole heaven, bound up in it. But in recognizing that life follows upon animation Aristotle recognizes that things that are alive are living beings. But Aristotle says that the soul is unchanging, thus denying it all natural changes. Plato also proves it to be self-changing as being changed by itself and changing bodies by changes of itself that are not natural but activities of the soul that are transitional and therefore initiated as changes.[100] For, to prove that Plato does not say that the soul is changed through natural change, listen to what is said in the tenth book of the *Laws*: (896E) 'Soul leads everything in the heavens ... by its own changes, of which the names are "wish", "forethought", "consideration", "planning", "opining", truly and falsely',[101] so that the soul also is unchangeable in the way in which it initiates change. Aristotle also follows Plato in saying that the first changer is unchangeable, the soul as unchangeable in the way that it initiates change, intellect as without transition in its activities. For since before the externally changeable there must be the self-changing as the immediate cause of the externally changed, and before that which is changed in any way that which is not changed, whether by another or by itself, the unchangeable is the cause also for the self-changing of its being and its changing, because the self-changing, being in a double state and exhibiting agency together with being acted on, is not sufficient to fill the role of a principle. This is so whether the soul is the first changer (it also is shown to be unchanged in every way in which it initiates change), or whether it is intellect (this is unchanged in respect of every transitional change). So there is no disagreement on this matter between Plato and Aristotle, except about the name of the self-changing, because Plato calls the soul self-changing but Aristotle says that the living thing initiates change through the soul which is unchanging in the way that it initiates change, but subject to change with respect to the body.[102] This is a consequence of Aristotle not thinking it right to call anything a change except those that are natural.

Therefore he does not think it right to call the transitional activities of the soul changes, while Plato, starting so to speak from the lowest level calls that which is unruly and disordered the first species of change, because according to a first manifestation of the forms generation comes about in a disordered way. The second species is that of natural objects, the third the transition of psychic activities, the fourth that of the intellect, which is without transition but arising to activity from its being.[103]

201a27 That of the potentially existent, when it is active in actuality [not as such but *qua* changeable,] is change.[104]

In what was previously said he both solved the objection brought against the first account of change and reminded us that natural objects exist not only potentially but also in actuality; he now gives a more complete definition by adding that that potential object whose actualization is change is not only potential, but it is completely something also in actuality, since it is a determinate nature among existents. Change is its actualization, not *qua* actually existing, for it remains as it is in that respect and rather initiates change, but is not changed or transformed.

But there are two different versions of this passage. According to Aspasius and Themistius and the majority of copies [it reads]: 'That of the potentially existent, when it is active in actuality, not as such but *qua* changeable, is change.'[105] But according to Alexander and Porphyry [it reads]: 'Indeed that of the potentially existent, when something actually existent is active, whether it itself or another, *qua* changeable, is change.'[106] Alexander knows the version of Aspasius as well, but prefers the latter, since it adds another difference in change. For the added 'whether itself or another' reveals that one sort of change is an activity from the thing itself through its potentiality, such as internally caused growth and diminution and natural travel in place and the voluntary moving about of animals.[107] Or change is externally caused as in the cases of things forcibly moved and all that comes about through a craft. The one is to be active, the other to be acted on. What is common to both cases is that the actualization is with regard to the potential, which is no less being acted on than acting. But it is worth investigating whether perhaps forced motion or coming to be through a craft is not natural change, yet now the discussion is about natural change, of which nature is the principle. For forced motion such as that upwards of a clod of earth is not through potentiality and natural disposition; otherwise it would not be forced, unless indeed there is also a natural potentiality and disposition to be forced. For not everything can be forced to become everything. But maybe the internally changed are not only changed

but are double and have within them that which initiates change and
that which is changed. As in the case of the motivated motion of
animals, the soul is clearly seen initiating change, even if itself
unchanged, while the body is changed. Similarly that which is growing and that which changes its place has one element that increases
it and transfers it, whether a nature that causes growth and change
of place or else something else which aids nature both in causing
growth and in repositioning, since nature also is not self-changing.[108]
If this view were victorious, in all cases what is changed is changed
by something else and not internally; and this is specially the view of
those who do not expect to see self-change in one and the same thing,
but divide it into the changer and the changed. And perhaps the
reading of Aspasius is safer; but if the addition according to Alexander
has also any merit, 'of itself' is not to be understood as referring to
one thing but to that which has the cause initiating change within it,
if there is also something in addition to the changed, such as the soul
and nature. This is obviously distinguished from that which is
changed externally by something else, as is seen in the case of things
which are changed forcibly or by a craft. But perhaps 'by something
else' is more appropriate in the case of things which are externally
but naturally changed, as air and fire come about from heated water
and in the case of things that are altered. However, I do not know
why Alexander thinks what follows better accords with the latter
reading than with the other. For I think that the internally initiated
appears nowhere in the examples, which are from the crafts, even if
health and disease are in some way natural.

201a29 I use the word '*qua*' in this way: [bronze is potentially a
statue, but, however, it is not the actualization of bronze *qua*
bronze that is change. For it is not the same thing to be bronze
and potentially something, since if it were simply and by definition the same the actualization of the bronze *qua* bronze would
be change. But, as was said, they are not the same thing. This
is clear from the case of opposites; for being capable of being
healthy and being capable of being diseased are different, for
otherwise being diseased and being healthy would be the same;
but the substrate which was the healthy and the diseased,]
whether moisture or blood, would be one and the same. [But
since it is not the same thing, just as colour and being visible
are not the same thing, it is plain that change is the actualization of the potential (*dunaton*) *qua* potential.][109]

He proves clearly by illustrations that what changes is changed
through the potentiality in itself, and that there is something which
is also actual in what changes, and that what is changed is not

changed in that respect, and also at the same time proves that even
when in substrate as well what is actual in some respect and what is
potential in another respect are the same thing, still these are in
definition different and not the same. Also he said 'if it were simply
the same' as equivalent to 'in all respects' in order that it should not
be only one in substrate but also in definition. He says that if this
were so an obvious absurdity would follow for one who said that the
actualization in accordance with potentiality of that which actually
existed was change, that the actuality of what actually existed was
change through its actuality, given that the actual and the potential
were the same in definition. This is absurd, because the actual, in so
far as actually existing, has its own activities without transformation.
For bronze, if it is active in some way as bronze, but is not potential
in any other respect, is active without transformation and without
change. But he seems by the rebuttal of the antecedent to rebut the
consequent by 'but as was said, they are not the same thing'. However,
that is not how he reached the conclusion, but this was a consequence
of the actualization of the bronze *qua* bronze being a change. But he
proves that the potential and the actual are not the same in definition
by the case of opposed potentialities in this way: what is actually
existent and a substrate to the opposed potentialities, as body is to
the capability of being healthy and the capability of being diseased,
is the same and not different; the potentialities such as the potentiality of being well and the potentiality of being sick are not the same
but different; therefore the actually existent is not the same as the
potentialities that it has, but different; for all things are either the
same or different. The question is in the second figure.[110] 'But as was
said they are not the same thing' reminds us of what was said about
this earlier. He said: 'Since some things are the same in both potentiality and actuality, but not simultaneously or in the same respect',[111]
also because 'it is not the same thing to be bronze and to be potentially
something'. He proves that the capabilities of being healthy and being
diseased are different from the fact that their actualities are different.
For having assumed for potentiality that as the actually opposed are
in respect of difference and identity so are also those that are so
potentially, he argues hypothetically in the second mode as follows,
converting by negation: if the potentially opposed are not different
but identical the actually opposed will also be so, which is absurd; for
the actually are plainly different, so that the potentially are so as well.
Also it is clear that the potential are the cause of the actual, since
they pre-exist; for they must be of a nature to act in that way. But the
actual are more obvious than the potential; that is why he proved it
of the potential from them, saying 'for being ill and being healthy
would be the same'.[112] For if the potentiality is the same so is the
actuality also, which can be plainly seen to be absurd. So if the

potentialities are different, but the substrate of the potentialities is the same, what is actually so would not be the same as the potentialities, which he set out to prove.

He called the substrate, which is what is healthy and ill, either moisture or blood since the object of the present inquiry was not to determine in what primarily health and illness reside, as some have done, whether it be in moisture and in juices primarily or in blood as what contains also the other juices, or in breath as according to the majority of ancient physicians, or in solids, or in the balance of the primary qualities. For there have been many opinions on this topic, some leaning to the most immediate objects, some to the intermediate, while some go back to the strictly primary, as did Hippocrates and his followers, who ran right back to the primary qualities of the primary elements.[113] Now the whole syllogism which he uses to prove that the substrate is one thing and the potentiality another is like this, being categorical in the second figure and plainly argued with reference to opposites: the substrate to opposed potentialities is one and not different; its potentialities are different from each other if its actualities are different; therefore the substrate which is actual is not the same as its potentialities; but if it is not the same it is different; for everything which is not the same is different. One may also draw the conclusion hypothetically in this way: if each of the potential opposites which are in one substrate is the same as the substrate they must be the same as each other; for things identical with the same thing must be identical with each other. But if it is absurd for opposites to be identical with each other, so also the protasis is absurd, that each of the opposites is identical with the substrate.

He proves that the potentiality is not the same as what has the potentiality in account, even if their substrate is the same, also through the example of colour and the visible. For of these the substrate is the same, but their accounts are different. For the colour existing in it is one thing, the potentiality which is being visible is another. For also colour is what changes the actually transparent, by which colours are seen, or it is the limit of the transparent *qua* transparent, as he himself defined it.[114] For sight, having traversed the transparent, meets with the colour. But what is visible is that which is capable of being seen; and that is an attribute of colour.[115] But the attribute is not the same as that of which it is an attribute. Therefore also those who define colour as the proper sense-object of sight define it by its attribute. For this is not the essence of colour, but of being visible, which is an attribute of colour.[116] It is clear that to be visible is not the same thing as being a colour also from the fact that the visible, even if it is also a colour, still is not received *qua* being a colour but through its capacity to be seen, but colour is not such through its being potentially colour, for that is not yet colour, but through its actuality. However, actual colour does not get its actuality *qua* being seen. For what is white exists in the same way by its own nature both seen and not seen. Also being visible is a relation, for it is related to what is capable of seeing, but colour is not a relation, but self-existent. Therefore colour is not visible *qua* being colour; rather being visible is its property, not its definition. They who define voice as the proper sense-object of hearing err in the same way. For that is the definition of the audible, which is an attribute of voice. It is not even the only thing that is audible since a noise is something else audible in addition to voice. They err in a similar way who declare voice to be reverberating air, as does Diogenes of Babylon;[117] for on that account voice is a body, since it is generically air, and they declare it to be what is affected, i.e. the reverberating air, instead of the affection, which is the beating. However, even if a voice is produced through the beating, it is not so through what is beaten. Also, as Alexander says, voice is 'the deliberate beating of breathed out air through the vocal organs'.[118] But perhaps it is not the beating, but the echo of the beating.

So if these things are true it is not the same thing to be bronze and to be potentially a statue, but the substrate is the same, the accounts are different. So change is well declared to be the actualization of what is capable of being changed *qua* being capable of being changed. And the objection from the fact that some of the things that are being changed are actual, since they can also be seen to initiate change, will no longer trouble us any more. For it has been demonstrated that even if the potential and the actual are the same thing in the substrate the change occurs through the potential. So '*qua*' is very properly added.[119]

201b5 So it is clear that this is what change is; [and it happens that something is changed then when the actualization occurs and neither earlier nor later. For each is capable of being actualized at one time and not at another, such as the buildable, and the actualization of the buildable *qua* buildable is the building process. For the actualization is either the housebuilding or the house. But when the house exists it is no longer buildable. But what is built is the buildable. So the actualization must be the building process. But building is a kind of change.] However, the same account fits in the case of other changes.

Now that he has said that change is the actualization of what potentially exists, and since the potential seems to be regarded as in incomplete readiness, whereas actuality is something complete, his project is to demonstrate in the case of the potentially buildable both that change is its actualization and that this is nothing other than housebuiding, and that housebuilding is change. But before setting out these proofs so much must again be prefaced, that the actualization of the potential is undergone, rather than being an activity. But he calls it actualization because through it it is actualized by what is active and changed by what changes it, and because its route is towards actuality, and because while formerly inactive and in mere readiness it is activated to actuality, and perhaps also because natural changes that come from within have in a way the same thing to initiate and suffer change, and for that reason such an affection seems to be active in a way. For primarily nature gives the body condition by being affected.[120] He also demonstrates that there is such an activity in the potential by saying 'For each is capable of being actualized at one time and not at another, such as the buildable'. Also it is clear that he is speaking of each changeable thing, and of things in which there is potentiality. For things which are always in change cannot be sometimes active or in change, sometimes not. For they are always changing, even if a different change at different times. This is also quite obvious, that what is potential must necessarily be transformed into actuality at some point if the potentiality is not to be pointless, and that this is necessarily so. So why did he say 'is capable'? Is it because many potentialities perish before being transformed into actuality and acting or being acted on in that respect? For many have perished before growing teeth or a beard or procreating though having a capacity to do so, and have not been actualized in respect of the potentiality. Perhaps it is also because things that are strictly potential, which are things that come and cease to be, are actualized contingently and not necessarily, even when they are active in respect of the potentiality. But it is sufficient for the argument, even if not all potentialities but only some are transformed into actuality. For in these change will be seen to occur.

So thus also the buildable which is potential may sometimes be actualized as buildable, when it is being built, sometimes not, for example if the stones and mortar are simply lying unused without yet being changed by the builder. As to what is the actualization in the case of the buildable, he says that 'the actualization of the buildable *qua* buildable is the building process'. He proves that it is this by a division. Either the potential itself before transformation is its actualization, e.g. the bricks and mortar are actualization through their potentiality, which is absurd, for they existed also before the actualization and in general the potential is one thing, its actualization another; or else what is actual is its own actualization, such as the house; but this he also proves impossible[121] as follows: the actualization of the buildable is when the buildable remains; therefore the actualization of the buildable is not the actualized house. So if the actualization of the buildable is neither before the building process nor after the building is completed, but the buildable is being built between it being potential and being actual, it is clear that the actualization of the buildable lies in its being built. But the actualization of the buildable that lies in its being built is the building process; so the actualization of the potential that remains potential is the building process. But it is quite obvious that building is a change, as are also walking and reading. So each of the presuppositions has been demonstrated. In the case of the buildable the conclusion is well drawn that the actualization of the buildable *qua* buildable is a change. And even if Aristotle omitted the conclusion as clear, having proved it in the one case of the buildable and the building process, that change is the actualization of that which potentially remains potential, he reasonably added that 'the same account fits in the case of other changes'. For the buildable and the building process differ in no way from other potentialities and their actualizations. So that every change is the actualization of the changeable *qua* changeable.

But Alexander notes that this reading is not contained in some copies, saying 'perhaps it is because some repudiate it through its unclarity and because it is said that what it says has already been said'. But it is quite obvious that what is contained in the definition of change has not been previously proved so plainly. It is still more obvious that if unclarities were to be struck out much of Aristotle would be struck out. However, it should be known that both Themistius and Porphyry explained these words as well when explaining the passage.[122]

It should be known that Plato says that all activity is change since he regards the activity of what is active from the aspect of the departure from being.[123] But Aristotle and his friends say that change is activity but not all activity is change. They do not so regard activity that is complete in itself; change is of what is incomplete *qua* incomplete, because it is of what is potential and not yet existent but about to be. Also it is clear that the difference is merely verbal, since the ones call 'change' only activity involving transformation, the others all extension from being.[124]

Chapter 2 begins here in modern editions.

201b16 It is clear that this is well said also from what others say about change, [and from the fact that it is not easy to define it otherwise. For one could not put change and transformation in any other genus, and this is clear if one looks at how some have placed it, saying that change is otherness and inequality and the non-existent. But none of these requires anything to be changed, whether things be other or unequal or non-existent. But also transformation is no more to these or from these than rather from their opposites. The reason why they assign change to these is that it seems to be something indefinite and the principles in the second column are indefinite because privative; for none of them is this, or of this sort, nor in the other categories. The reason why change seems to be something indefinite is that it cannot be classed among things potential nor among things actual. For what is potentially a quantity does not necessarily change, and nor does an actual quantity. Change seems to be an actualization, but an incomplete one, and the reason is that the potential is incomplete in respect to its actuality. That is why it is difficult to find out what it is. One must either treat it as a privation or a potentiality or as a simple actuality, but it can be seen that none of these is possible. So the characterization stated remains, that it is a sort of actualization, but such a one as we said,] difficult to see but capable of occurring.

Having stated his own view he also turns to an examination of those of his predecessors about change. Alexander also interprets the other things that are said here in a way that I do not understand as well as what is said at the beginning, that 'it is clear that this is well said also from what others say about change' and that it is impossible to put change in any other genus. He says: 'He confirms it from the opinion of others who by placing change outside such an activity said nothing sound about it.' But perhaps it is more wise to understand Aristotle as criticizing the more general sense of the account by the ancients; while he also shows from their ideas that he himself had well defined change. For they as well place it in otherness, inequality and not-being because change seems to be something indefinite. For the Pythagoreans have two columns, in one of which is change; this one contains the principles that are privative and indefinite, among them otherness, inequality and not-being (for they accepted the ten oppositions as principles).[125] So change reasonably seems to be indefinite to them also, as it does to us. The difference is that to them it seems to lie in the column of the privative and indefinite through its opposition to the definite state of being static, but to us because it is not possible to assign it either to potentiality or to actuality. For what is only potential is not yet changing and the actual is no longer changing. So if it is neither the potential nor the actual, nor even, what it especially seems to be, activity (for change is an incomplete activity, because the potential of which change is the actualization is incomplete), change reasonably seems to be indefinite.

He said that the principles in the second column of opposites were privative, referring obviously to the ten which they include in it because according to him the principles of opposites are form and privation. The ten principles in the superior column will fall under form, those in the inferior under privation. But perhaps before him they also called the one formal the other privative. Aristotle himself recounts the ten members of the columns elsewhere:

good	bad
limit	limitless
odd	even
one	many
right	left
light	darkness
male	female
resting	changing
straight	bent
square	oblong

Those in the second column are also called privative, not when they are viewed as forms but when viewed as falling away. Therefore they are neither this nor of this sort. For privations, even if they be in the same categories, still are so as by-product.[126]

Also I myself believe that he gives an account of these as relating to the statement: 'It is clear that this is well said also from what others say about change.' But Alexander says that he gives no account of anything consequent on this but he says that it is concluded from the second statement that 'it is not easy to define it otherwise' that 'also one could not put change and transformation into any other genus'. Also he accuses the text of being unclear and inconsequent because he did not add 'nothing of what is said about it by the others is sound' to the next statement. 'For', he says, 'if that is said first what he adds follows on, which is from "it is clear if one looks at it that some of them place it as otherness" to "it is hard to see but is capable of existing".' He also understands the statement in another way, and says: 'or perhaps it is better to take "also one could not put change and transformation into any other genus" as being said as equivalent to "in any other natural kind" and "it is clear if one looks at it that some of them place it as otherness" would follow on it.' He also says that this passage is not contained in many copies and has not been commented on for that reason. As it seems to me, he himself made this comment because his exposition was confused.

He [Alexander] made many objections on the ground that in Plato change is said to be otherness, of which I think the following to be the most important; or perhaps they are all, for they are interconnected. He says: 'He means that change is either otherness in relation to something else, or internal otherness, or becoming other and being transformed. But neither is what is other in relation to something else always changed, nor is that which has internal otherness, such as heat and whiteness, since neither the potential nor all the actual is always changed. But if by otherness they mean becoming other and being transformed they do nothing other than say that change is change. 'Also', he says, 'Plato says that change is that which is not in the *Sophist*'.[127] Leaving aside what Plato plainly said, he tries to prove this by an argument. For since he said that of existents some changed and some were constant, he took the difference in being to be change and constancy.[128] But if these are the differences of existence, but a genus is not categorized by specific differences, then nor would the being of change and constancy be so categorized; but, if this is so, then they are not existents. Also it is clear that even if they are not so categorized as by their species then the genera still exist by their differences. For the rational is an animal, and so is the non-rational, even if animals are not so categorized. So thus both change and constancy, if they were different sort of existent, would not be outside existence. But Plato plainly says that change is non-existent where he says 'therefore of necessity there is not being in the case of change and with respect to all its kinds'.[129] But he says that it has not being not as existing nowhere in any way, which also is strictly unnameable, but because of its otherness with relation to being, which he says is one of the genera conceived in its distinct individuality, but not with respect to the content of all existents. As he says that change is non-existent through its otherness in relation to such a being, so he also says that it exists through its participation in being. He speaks as follows: 'So clearly change is really non-existent and existent, since it participates in being.'[130] Having said that the Pythagoreans also were of this opinion that change was not-being, he wisely offers a defence on their behalf, saying that perhaps they did not say this flatly nor as holding absolutely that it did not exist, but on the ground that what was being changed was not yet actually that into which it was being changed but was still in an incomplete state; for the potential is like that.[131] That is the view about change of Aristotle also. 'As for these', he says, 'otherness and inequality and the non-existent, if Plato and the Pythagoreans held that they were causes of change, that was possible, but not sufficient for the account of change. For what causes is not the same as what is caused. Also those who said it was non-existent, even if they spoke some truth, still gave an attribute of change and not its definition.'

It will soon be clear that Plato mentioned inequality as a cause, when we provide Plato's own words. But for the present let it be known that Eudemus also, when before Alexander he gives an account of Plato's opinion about change and attacks it writes as follows:[132] 'Plato calls change the great and the small, the irregular and the non-existent, and whatever has the same tendency as these. But it is plainly absurd to say that this is change. For it is when change is present that that in which it is seems to change. But it is laughable to say that when something is unequal or irregular it has got to change; it is better to say that these are causes, as Archytas did.'[133] Also a little later he says: 'The Pythagoreans and Plato do well to count change as indefinite (for indeed nobody else has spoken about it); for it is in fact not definite,[134] and is the incomplete and the non-existent. For it is coming to be and what is coming to be does not exist.'

Aristotle also examines the account 'saying that change is otherness and inequality and the non-existent. For', he says, 'the other and the unequal and the non-existent ought to change straightaway.' But now it can be seen that nothing changes as being such. 'For', he says, 'all things that exist are different from each other, and yet not all things are changing.' In the case of the non-existent it is even more absurd, if the non-existent changes. But if somebody were to say that things undergoing change either change into these or from these, and that is why change seems to be these things, he anticipates that sort of objection and says 'change is no more to these or from these than from their opposites', and obviously to their opposites, e.g. from identity and equality and the existent, and to these. 'But', Alexander says, 'it is possible that it is said that change comes about from otherness and to otherness no more than from and to identity because even if what is changed chiefly changes from other to other, still the things that are changing themselves, such as what is becoming white or becoming black are generically the same as each other, having qualities and being identical in that respect. Similarly the subjects of growth and diminution have quantities, and so in the other categories.' But Alexander, when he has said that the reason why they thought change to be indefinite was that one could not place it in any of the definite kinds, since of things that exist and are definite some are potential, others actual, and change is among neither of these, having only just made this wise judgement revokes it by adding: 'As I said, by these considerations he also shows that it is not easy to define change otherwise than as he did.' However, through this Aristotle shows that those who said it was indefinite were aiming at the conception about change, and he well shows at the same time the reason for it being indefinite and for the difficulty of the discussion about it, at one time saying that it is not the potential nor the actual nor the complete actuality within which all things are contained, at another time saying that it is necessary to assign it either to privation, since conditions are stable, or to potentiality, since the actual is unchanging, or to actuality, since the actual seems to be what remains.[135] But not simply to privation; for privation is of form. But of what form is change the privation? If of the form of rest, why is not rest rather the privation of change? Nor is it potentiality. For potentiality and the potential are unchanged as such. Nor very surely is it simple and complete actuality, but incomplete, because it is of the potential, which is incomplete.

For this reason it is difficult to see, since knowledge is of the complete, but yet it is not impossible for it to occur. If change is that sort of thing Aristotle did well not to place it among the kinds of being, as did Plotinus.[136] For the kinds, i.e. the categories, are of things complete and definite, but change is viewed in all kinds, through the transformation from potentiality into actuality.

But next it is time for me to set out the statement of Plato written in the *Timaeus*,[137] against which Aristotle and Aristotle's commentators seem to have spoken: 'Change is never willing to occur in uniformity. For it is difficult for there to be anything that will be changed without a changer, or that will initiate change without what will be changed; or, rather, it is impossible. But there is no change without them. But it is impossible for these ever to be uniform. So we should always place stability in uniformity, change in variation. Inequality again is the explanation of the nature of variation, and we have discussed the origin of inequality. But we did not say how at some time things not separated in kind have each ceased their motion and travel through each other.' So, having said how through the concentration of the universe upon itself the more rare parts enter into the denser and separate them and are themselves congregated together, he draws his statements to a conclusion by saying: 'Thus through these the coming about of variation is for ever preserved and provides the everlasting change of these things, present and future.'[138]

It is clear to all that Plato made no reference to otherness in this passage. But even if Plato or the Pythagoreans say that there are these three, otherness itself regarded as a form, the other through partaking in otherness, and becoming other, when they said that becoming other was change they meant that sort which was alteration and activity. But Aristotle switched to the otherness which is form, by participation in which things became other. Thus he reduced the account to absurdity by saying that 'it is not necessary for any of these to change, whether they be other or unequal or non-existent'. For, if otherness were change, things which participate in otherness would change at once, these being the other. Also one should not accuse Aristotle of frivolously also not choosing to take otherness which is as form as being change according to them, but the other, and thus reducing the account to absurdity when he says that it is not necessary for things to change even if they are other. For he did not introduce the other into the account as being change but as participating in otherness, and, if that were change, the other, as participating in change, would inevitably have to change.

Eudemus also set out more clearly Aristotle's proof in the passage I recently quoted (431,10), by: 'For it seems that it is when change is present that that in which it is is changed; but it is laughable that if this is unequal or variable it is obliged to change.'[139] Similarly it is not plausible to take change as being either inequality or non-existence or what partakes of these, but rather as becoming unequal, and coming into existence or ceasing to exist. For these are changes, certain transformations of a special sort in each kind from potentiality to actuality. For they do not call what is already distinguished and remains in its species otherness. Also inequality and the variation of the inferior in relation to the superior or of the changed in relation to the changer, which seeks to bring the incomplete to completion and implants actuality into the potential, assigns the first cause of change as coming from the superior.[140] One can come to understand inequality, variation and otherness also from the dissimilarity of the changer to the changed. For the changer is one thing, the changed another, and these are unequal and vary. For these reasons change is constituted around what is changed by the changer, being introduced through the variation itself, not as being bounded in respect of the variation but in respect of an indefinite progress in the variation towards the superior. For a mixing of the definite and the indefinite constitutes change. For the variation and difference of the changer with respect to the changed, of actuality with respect to potentiality, of the first to the middle and the end, comes to be in them as the origin of transformation, since all of these are differently disposed towards all in the case of change. For so long as something remains one and the same it is not of a nature to change. So reasonably the other and the unequal and the non-existent, which are already established in that thing, are not changed, but those things that are becoming other and becoming unequal and coming into being, these are changed as they seek on each occasion to become different and other.

Even if transformation to the unequal from the equal does come about in respect to being indefinite, the transformation comes about through the equal and through becoming unequal and through departure from equality. For to the extent that the quality of the equal is stable a thing is not changed either from or to equality. Again, the transformation from or to something is not viewed singly, but with regard to the combination of that from something together with that to something. Also the actualization with respect to becoming other combines the two. But, indeed, the combination of the extremes is a containment of things unlike and different. So long as that is present what partakes in it must be changed. So in that way change, which is indefinite, is assigned to the privative. The varying is the privative of the stable, the unequal of the equal the other of the identical, non-existence of existence. So change is reasonably seen as in them. For in so far as they are deprived of form and aim at it and speed to attain to it, things changed change to that extent. So the Pythagoreans reasonably assign change to privation;[141] and they say that it is indefinite as being in itself neither matter nor form nor potentiality nor actuality, but a mixture of these and a mean between them so as to be none of them. And non-existence belongs to it as not existing either potentially or actually but as progressing from the one to the other. Also the incompleteness in the activity concerning it is of a sort such that it does not yet achieve actuality and form. Also its contact with all of them and its being a mean in the whole of them is of such a nature that it is neither form nor privation nor potentiality nor actuality completely. There is an actualization in regard to all these which is hard to capture, separate from all of them and in another way occurring with all of them, present in the potential so long as there is potentiality, but no longer existing in any way when the potentiality ends. But, if it is in neither potentiality nor actuality, how can change be said to be actual? Or is it not simply activity but incomplete activity, not as incomplete transformation but as incomplete actualization and incomplete form? Change is different from both the actual and complete actuality since in their case the potential has ceased; but in the case of change the actualization of change is completed while the potential remains.

But we must go on to what comes next.

202a2 Every changer of the sort mentioned is also changed [if it is potentially changeable and if its lack of change is rest; for the lack of change in that to which change occurs is rest. For to act on a thing in a certain way is itself to initiate change. This is done by contact,] so that it is also affected at the same time.

He has previously said that that which initiates change naturally is also itself changeable.[142] For everything of that sort is changed as it initiates change. Now he wishes both to provide clear examples of things reciprocally changed in initiating it and at the same time, as Alexander says, to separate off the divine body from being reciprocally changed and affected. By one example he seems to separate that which initiates change and is reciprocally changed naturally in coming to be from that which initiates change without changing, as does the first and intellective substance, and from that which initiates change as always changing, such as the heavenly bodies.[143] For what initiates change naturally initiates change as potentially changeable, which is why it is also reciprocally changed and also not only undergoes change potentially but also initiates it thus. But the first substance is altogether free from potentiality both of being changed and of initiating it. Also it is not changeable, being entirely unchangeable and it does not initiate change potentially. For it transcends all incompleteness and all potentiality. So it will never potentially initiate change, nor will it sometimes do so, at other times not, nor in some things and not in others. So the natural changer is distinguished in this way from the unchanging changer, and yet more by the fact that, even if what initiates change naturally sometimes remains unchanged and potentially changeable, its lack of change is not the same as that of the first substance. However, in the case of those that do so naturally, their lack of change is that of what sometimes is changed and, he says, is called rest for that reason. 'For the lack of change in that to which change occurs is rest.' But in the case of the first substance which is altogether unchanging its lack of change is not rest but self-identity and an everlastingly untransformable constitution.[144]

He would also distinguish things that come to be and cease to be from the heavenly bodies by the same examples, for they also initiate change without being reciprocally changed. For even if they also are changed by the unchanging causes, still they are not changed by the things that are changed by them. Nor are they potentially changeable, since they are in everlasting change, unless one takes on some potentiality towards one change and another, since they move at different times in different places. But they also do not rest, just as the first substances do not, except that the latter do not rest because they are not of a nature to be changed. 'For the lack of change in that to which change occurs is rest.' But the former do not rest, because they are ever-changing. But I think that he proves that the lack of change in that to which change occurs is rest and not the changelessness of the wholly unchanging by the fact that to initiate change is the action of the changer on the changeable *qua* changeable, i.e. on what sometimes before being acted on is only potentially unchanging and at rest in respect to the change that it is about to undergo. For what is at rest is unchanging as being something that sometimes is changed. But what so acts on it is natural and corporeal and changes by contact because the qualities of doing and undergoing in these bodies are present in the bodies which are their substrate and act inseparably from them. Having also this potentiality in them they act as actual and suffer through their potentiality. That is why what so initiates change is also reciprocally changed. For they are of the same matter and are opposed under the same proximate kind. And since each thing has one of the opposites actually, what is actually hot but potentially cold warms but the actually cold but potentially warm is warmed by it. This is why the latter, in so far as it is actually cold and potentially hot, also reciprocally conveys cold when it is warmed, and what is actually cold but potentially hot is warmed by it. So if change is stated to be the actualization of the potential *qua* potential, but is actualized by what actually touches it and also has some potentiality itself, it is clear that what so acts also is acted on and the changer is reciprocally changed. It is also possible to set out the argument syllogistically as follows: that which naturally or in becoming initiates change, because it is not always actually what it is potentially, is changeable, and because it is corporeal it changes by contact; but what is potentially changeable and initiates change by contact is also reciprocally acted on and changed, the less dense separating out the more dense and being collected together by it, and the denser conversely.

It should be known that Aspasius writes the text as follows: 'every changer such as described is also changed,[145] being the potential initiator of change'. And what is soon added chimes in with this thus: 'for to be active towards this, as such, is what to initiate change is; this it does by contact so that it is acted on as well at the same time'. For being the potential initiator of change and initiating change by contact, when it initiates change, it is also reciprocally changed. For things that come to be and cease to be are such. Alexander knows this reading as well and gives a useful account of it as being accepted by others, but he puts first the other which says 'every changer such as described is also changed, being potentially changeable'. Also this is clearer by distinguishing which sort of changers are reciprocally changed, because not all are. For neither the altogether unchangeable nor those things that are always in actual change, but only those that potentially initiate change, are also themselves potentially changeable.

Having said this he again inferred the definition of change from it by 'therefore change is the actualization of the changeable *qua* changeable'. 'And it seems', says Alexander, 'that he now says more clearly what change is than when he said that it was the actualization of the potential *qua* potential.[146] For, indeed, while the potential is in all the categories not all the actualization of the potential *qua* potential is change. At any rate relations are potential in so far as potentiality becomes actuality, as for example the potentially double becomes actual and there is no necessity for it to be changed, but it does so through that of which it is double being set beside it. In this way it becomes actual from being potential, but it certainly does not change. On the other hand everything, if it is changed in some way, thus comes to actuality from potentiality. In this respect the actualization of the potential *qua* potential is more extensive than is change.'

After this Alexander well insists on what I earlier quoted:[147] 'Also in the case of relation the potentially double is not changed. However the potentially half was changed and became actual, in order that the other should become double. But that of the potentially double and the potentially half were not different potentialities. So, with respect to the actualization of the potentiality, change in their case also was of one potentiality with respect to the substrate, which brought about the different relationships of the related terms.' However, while Alexander wrote this in these words, I think it right to call to mind what was previously said here,[148] when I said that there was change also in the category of relation and not only in the four categories. For in the above Alexander clearly agrees that even if one of the relata be changed the other is changed as well, since the relationship is one. But Themistius here also judges[149] that there is not change in the category of relation because actualization in relation does not preserve the potential, but the transformation from potentiality to actuality occurs all at once. For what is on the right does not become so gradually but all at once. 'So', he says, 'it is necessary first to demonstrate incomplete actualization and thus to examine the definition.'[150] But if, on the one hand, the transformation from potentiality to actuality is instantaneous, what is absurd about change in the category of relation being instantaneous? If, on the other hand, things in becoming come about in time and time is involved with them change has not been lost through this. 'So', says Alexander, 'either he exchanged "potential" for "changed" as being clearer or, if somebody insisted that the potential was wider as being in all the categories, he will reasonably seem to have moved over to the more specific account of change, calling it the actualization of the changeable *qua* changeable.' Alexander also mentions another advantage of the exchange into 'the changeable'. 'For since he wants to show that change is in the thing changed and also that the changer which is naturally a potential source of change becomes actually such when it acts on the changeable but is not being changed, nor is change in it, he reasonably took the thing changed as what change is in, and not the changer through which change comes about in the changeable. So he made the definition to be of equal extension by converting "the potential" into "the changeable".'

Some also criticize the definition of change because it makes use of the changeable, which is equally obscure as change and is recognized through change.[151] But Alexander defends it, saying that 'if change is a relation, since change is of the changeable, and if to be something relative is the same thing as to be in some state towards something, he reasonably in defining change made use of the changeable to which it is essentially related. Also, in general, recognition of change does not depend on the changeable but takes into account the changeable only so far as change is the actualization of what is of such a nature as to be changed in accordance with its nature and in no other respect. For the buildable, when being actualized in accordance with its nature is built while its nature remains. Thus what is changed is changed altogether.'[152] But perhaps, after the multiform unfolding of the definition, there is nothing strange in his carrying it forward in this way into a clearer argument by putting in 'the changeable' instead of 'the potential'. He says that this actualization comes about by the touch of the changed by the changer. In the case of things that are naturally and bodily changed he says that the changer and the changed touch each other, whereas in the case of change being initiated by the incorporeal he says that the changer, through being incorporeal, touches the changed, using 'touch' in the case of the incorporeal in a wider sense, in the way that we say that abuse touched us; but we do not also say that the changed body touched the changer, as we also do not say that we touched the abuse. But if the changer is reciprocally changed in so far as it is potential, by this he also shows more conclusively that change is the actualization of the changeable *qua* changeable.

202a9 The changer will always bear some form [as a substance or a quality or a quantity, which will be the origin and cause of the change, when it initiates change, as the actual man] makes a man out of what is potentially a man.

He has shown that what is changed, so long as it is being changed, retains its incompleteness and potentiality; reasonably, he further indicates concerning the changer that this must be complete and have a form. For things that initiate change, and produce naturally and immediately, produce things like themselves and bring the things changed by themselves into their own form. So if the changed is changed in substance the changer must be a substance, if in quality a quality, if in quantity a quantity, 'as the actual man makes a man out of what is potentially a man'. For the changer, being complete, is the initiating origin and the efficient cause of the change. But how is this so in the case of what initiates change of place? For what initiates change of place will be neither a place nor in a place, as substance initiates substantial change and quality qualitative change. Rather it is necessary in these cases that the changer should be complete and surpass the things changed, but surely not also in place.[153] At any rate the soul changes the body by change of place, but surely not as being in a place. Perhaps this is why Aristotle, more cautiously, did not say that the changer will bear a form such as the changed also has, but 'as a substance or a quality or a quantity'. So therefore perhaps there does not always have to be the same form implanted in what is becoming by the agent, but a suitable form in each. For also the builder builds according to his conception of a house, and the soul initiates change of place not through some chance cause within itself but according to the cause of change of place within itself which has the same form as is in that change.

Chapter 3 begins here in modern editions.

202a13 So the answer to our problem is plain, [that change is in the thing changed; for its actualization is by the changer. Also the activity of the changer is not something different; for it must be the actualization of both. For it is a changer by potentiality but actually changes by activity, while it actualizes the thing changed, so that the activity of both is one just as the interval from one to two and from two to one is the same and also upwards and downwards. For these are one though their account is not one. It is the same] also in the case of the changer and the changed.

After providing the definition of change and making the appropriate analysis of its parts he next turns to the second of the problems about change, inquiring and explaining in what change occurs, whether in the changer or in the changed, or in the changed from the changer, which is the relationship of things related. By that means he also distinguished the relationship of the thing related. For, since change is an attribute, it is altogether necessary for it to be in some substance. So, since 'change' appears in the names of both the changer and the changed, it seemed worth investigation whether change is in the changer or in the changed or in both. Since the changer is active and the changed passive, if change be divided into action and passion, the agency is in the agent and the passion in the passive. So he says that also the problem about these is solved by the definition of change. For it is plain that change is in the thing changed since it is the actualization of the changed by the changer. Eudemus says[154] that 'sense-perception implicitly contradicts those who say that change is to be situated in the passive subject of change; for we say "sees", "is seen" and "seeing", and "hears", "is heard" and "hearing"; so seeing seems to be in the seer and hearing in the hearer. It is the same with taste and the other senses. But these seem to be in the column of the agent and changer. 'We are deceived', he says, 'by being guided by language and not observing that it is the perceiver that is affected.' But why is there not also activity of the changer? For, if this is a preceding source of change and thus a potential changer, it is necessary that when it becomes an actual changer it should then initiate change and be active in respect of doing so. Why then is the activity of the thing changed change rather than that of the changer? But perhaps the activity of the source of change is not a different one; for the source of change is what activates the thing changed.

So the activity of both is one, being called more strictly an activity as from the changer, but in the changed it is an affection rather than an activity. But perhaps this is also called an activity in so far as, even though it is externally initiated, still, because of its internal nature, it is able to be changed according to its own suitability. So as the interval from one to two and from two to one is one and upwards and downwards are one in substrate, but two in definition, the ratio from one to two being a half and from two to one double, so is the relation of change in the case of the changer and the changed, since they are one in substrate, but is viewed one way in the changer, another way in the changed. It is clear also that when the source of change is transformed from being potential to being actual by what is actual, then the change is present in it as being changed by the changer and is a different change from that by which it, already actualized and become complete, changes what is changed by it.

So the activity and the affection are one in substrate, beginning from the actual and getting its form thence, but implanted in what is acted on, and having this character through the giving of the form by the agent like a breath into another, so that it is neither separated from the agent nor in it. It is from the agent but not excised from being in it. So if we were to start from the agent and initiator, change is to act and to initiate change, but if from the affected and changed it is to be acted on and be changed.

It should be known that in this area most write this passage more clearly as follows: 'for it is the actualization of this by the source of change'; but Andronicus writes as follows: 'for it is the actualization of the changed and from it'. He gives the interpretation that, even if the changer be external, the changed seems to be changed by itself, since it is brought to actuality through its inherent potentiality.[155]

> **202a21** There is a verbal difficulty; [for perhaps it seems necessary for there to be an activity of the agent and of what is affected. The one is agency, the other being affected, and the fulfilment of the one is a product, of the other an affection. So since both are changes, if they are different, in what do they happen? For either both are in the thing affected and changed, or agency is in the agent, affection in the affected. But if that too has to be called agency] it would be ambiguously so.

He has first said what he himself believes about change, that it is one, coming from the changer and in the changed; next he sets out a problem about this which bothers those who do not accept what he said. He calls this verbal (*logikên*) either because it arises from accepted opinions[156] or because its plausibility arises only from the words and is not supported by the facts. For it is in this way that the arguments of Zeno which plausibly abolish motion are called verbal. Or else he calls verbal what is more general and neither relevant or proper to the topic and which does not arise from relevant principles. For an argument from relevant principles is not verbal, such as that which proves that the soul is immortal from its being self-changing;[157] for it designs to proceed from the essence of the soul, as Alexander says. But the argument proving immortality from opposites coming from opposites is verbal;[158] for it does not proceed from related and proximate premises about the soul, but through general ones through which something else also might be proved to be such which was not of the same sort as the soul. For even if there needs to be a universal premise in demonstrations, still it should not be general, when we prove particular cases, in such a way as to be suited to matters of a different kind. Alexander well supplied suitable examples. But I make this additional point for the sake of lovers of truth, that the argument from opposites in the *Phaedo* does not purport to demonstrate the immortality of the soul in that way, but only that the soul pre-exists life HERE, and survives it.[159]

He expounds the problem by first making the distinction that there is some activity both of the changer and of the changed, or of the agent and what is affected, so that there are two activities. What he previously took as obvious, when he said that there must be an actualization of both, he now gets from the difference of the names which the activities have. For if the former is called agency and the latter affection these should be two different activities. And if their products and goals are in the one case a product, in the other an affection, these are different from each other. But where the products and goals are different, there the activities which have the goals are different. So having made these distinctions he next begins the division which dissolves the problem in the following way: if both are activities, that of the changer and that of the changed, these are either completely different or not different. If they are different, then since changes are attributes they are inevitably in some substance. So in what substance[160] are the two? For either both are in the affected and produced, or the changed, or else agency is in the agent and affection in the affected. But if somebody should wish to call also being affected agency, because it too is called an activity, being the actualization of the changeable, let him so call it, but let him know that this is called an agency in a different sense from that of the agent. The difference remains, even if the words are the same.

He left out the third section of the division which says 'or both are in the changer' both as being obviously absurd, since it results in the changed not being changed if it has no change within itself, while the changer has two changes and is changed with a double change, but also because the same absurdities come about for this section of the division as well, and even more so, as for that which placed both the changes in the changed. He will in any case prove as he goes on that it is absurd even to say that the one of the changes which seems related to it is in the changer, and if that is absurd it is much more absurd to place both of the changes in it. So those are the reasons why he omitted that section of the division as it is written in the majority of copies. But in some there is this addition written as follows: 'for either both are in the affected and produced or in the agent and subject, or else the agency is in the agent and the affection in the affected'. The addition is superfluous, since Aristotle raises no objection to that view.

But there is also a fourth section to the division which is opposed to that stated second, which says 'or the agency is in the affected, the affection in the agent', which he left out as obviously absurd. But the persuasive Themistius is of the opinion that 'but if one must call this also agency it would be ambiguous' was said as pointing to this section, as if, after saying that the agency is in the agent, the affection in the affected, he had added: 'once again it is not possible for such a thing as agency to be in the affected, unless somebody should wish ambiguously to call the activity of the affected agency'.[161] However, if it were said for this reason it would be necessary also for affection to be ambiguously said to be in the agent. This was not said as being manifest.

202a28 However, if that is so, [change will be in the changer; for the same argument applies to the changer and to the changed. So the result will be that either every changer will be changed, or will contain change but not be changed. But if both are in the changed and affected, agency and affection, and both teaching and learning, which are two, in the learner, then first the actualization of each will not be present in each, and then it is absurd for it to undergo two changes at once. For what will be the two alterations of the one thing] and into one and the same form? It is impossible.

He first deals with the section which has just been set out, which says: 'agency in the agent, and affection in the affected'. He takes it that, according to this hypothesis, just as the changed has its own change, so the changer has its own change, (for the same argument applies to both) and he infers that on this hypothesis one of two absurdities must be accepted. For either (1) every changer will be changed as well, since the changer contains change; but what contains change is changed. But it is absurd that every changer should be changed and it is not Aristotle's opinion, since he believes that the first changer is unchanging.[162] Nor does it accord with the facts; for there are many changers

Translation

that remain unchanged such as colour, beauty and good form. Or (2) otherwise it has change in itself but does not change, which is the greatest nonsense of all like that which possesses whiteness not being whitened. For that in which change is present is inevitably changed and what is changed is changed through the presence of change.

Having exposed this absurdity, he moved on to the other section of the division which was stated first, and next presents the absurdities for those who suppose that both are in the changed and affected – agency and affection, teaching and learning. The first absurdity is that the activity of each should not be in that which actualizes it but in something else, if the activity of the agent is to be in the changed. The second absurdity is that the same thing should be changed in one form by two different and opposite changes. For if the changer alters, let us say, and the changed is altered, it is not possible for the changed to alter in the respect in which it is altered. For since the activity is different its results must be different, just as it is not possible for a thing to increase in the respect in which it brings about alteration. For a thing is increased by what increases it and altered by what alters it. He harshly added an extra absurdity for this hypothesis – 'for what will be the two alterations both of one thing changed and leading to one form?' For it is still more absurd that also the changes that are two, which he called alterations, should lead to one form. It is clear that the activity of the changer and the changed ends in a single form from the fact that when the changer and the changed cease their activity the changed will be seen to have suffered a single transformation, so that it was to this that it was also changed. So it is impossible for them both to be in the changed, since the change comes about in a single form – heat or whiteness or something similar – and impossible for the two, being opposite, to be present together in the same thing, and further for the two, being opposite, to end in the same form. For when one activity of both is supposed to be in the substrate nothing prevents them being two in account and to end at the same form, but if two are supposed the hypothesis is impossible since there is one form. For even if one were to accept the hypothesis, the consequence that they end in a single form is impossible.

'But it is possible', says Alexander, 'to read the text in another way by giving *tines* an oxytone and not a barytone accent.'[163] For if the changes are two at once there will be some – *tines* – two simultaneous and opposing changes of the same thing and into a single form', and these are impossibilities. 'But', he says, 'it is possible to make the text clearer by transposition as follows: "then it is absurd for a thing to undergo two changes simultaneously and into the same form". And this is not only absurd but also impossible. Then, after this, "for what will be the alterations of the single thing?"; for these things are impossible because it is not possible to discover what will be the alterations of the single thing. He also added a special indication of the absurdity for each of the absurdities', Alexander says, 'and the absurdities are undergoing two changes at once and that the two should be into a single form. For he added to "it is absurd for it to undergo two changes at once" "for what will they be?" For this is absurd because there is no way to find out what they will be. But he added "but it is impossible" as well to "and two into a single form". For that is impossible. But why is it impossible? Is it because every transformation is from opposite to opposite and one is opposite to one? If so, it is impossible for the two changes, since they are two transformations, and clearly from two states (for there would not be two from one) to terminate in the opposite. But the impossibility will be still more visible if we bear in mind that also the two changes of the changer and the changed, or the agent and the affected, will be at once in the same thing, being both opposite to each other and in that only into which the transformation occurs. But it is impossible for opposites to hold simultaneously with regard to the same thing.'

But perhaps Aristotle thought that there being two changes of the same thing into a single form was sufficient for the uttermost absurdity and so did not recall that they were opposites. Therefore he instanced two absurdities – that there were absolutely two of the same form and that the two which were different should lead to one end, even without their being opposite. But the opposition increases the absurdity. But Alexander allows here that two should be opposite to one in the case of excess and deficiency alone. For these seem to be opposites both to the intermediate and to each other. But it is worth noticing that both are opposite as one to the intermediate; for it is as the incommensurate are to the commensurate, and as the unequal to the equal. So the principle saying that one is opposite to one is not infringed by this case.

Having stated the absurdities that arise for those who place both changes in the changed and affected, Aristotle made the observation that the same absurdities will arise also for those who place both in the changer and agent, and still more so, since change is in the changed. That is why he omitted that section of the division; 'for the changed will be changed without changing.'

202a36 But the actualization will be one. [But it is illogical that there should be one and the same actualization of two things different in species. Also, if teaching and learning are the same, and agency and affection, then to teach and to learn will be the same and to act and to be acted on, so that it will be necessary that the teacher should be learning everything] and the agent should be being acted on.

The beginning was a division, that either changes are different or not different but, obviously, one, which he indicated by 'if different, in what are they?',[164] so that 'if they are not different' needs to be added. So having first set out the absurd consequences for those who say that they are different, he next turned to those who supposed them to be one. He states the absurdities that seem to follow for this view also, which he will also soon abolish as he explains the problem away by a distinction.

What then is the absurdity that follows for those who say that acting and being acted on are one change, e.g. teaching and learning? He says that it is that there should be one and the same actualization of two things different in species. It is clear that the changer and the changed are different in species, since the agent is actually what it is, while what is acted on is potentially of that sort, but actually is the opposite of what it is potentially, since transformations are from opposites. So they are not only different but opposites. So as it is absurd to say that the actualization of white and black,[165] which differ in species, is the same, as the one enlarges the iris, the other concentrates it, so, since the agent and the affected are different in species, a still more manifest absurdity follows on the absurdity that there should be a single actualization of two things different in species, since each species ought to have its own actualization. For if teaching and learning and agency and affection are one and the same then to teach and to learn will be the same. You will see how the absurdity is still more obvious if you take the teacher himself instead, so that the teacher will be learning everything that he teaches and the agent should undergo everything that he does.[166] For of these three, teaching, to teach and the teacher, teaching is the activity itself, to teach displays the essence of teaching in actuality, and the teacher its being together with the activity. So the absurdity is apparent both in teaching, still more so in doing it through the exhibition of the essence which contains the opposition, and still more obviously than in these in the essence itself. So since by the arguments stated both have been refuted – that the activities should be simply two and simply one – it will be proved that in a way there is one and in a way there are two, one in substrate, two in account.

202b5 But perhaps it is not absurd for the actualization of one thing [to be in another; for teaching is an actualization of that which can teach, but it is in something and not cut off,] but of something in something else.

From now on he begins to remove the difficulties, not all of them, but such as attack the true opinion which says that change is one in substrate, but two in account. Therefore he dissolves all the problems about change being one which he set out later, but not all of those about changes being double, but only such as can conflict also with there being one which is also in the changed. This he did reasonably; for it is these problems which particularly affected the definition that says that change is the actualization of the changeable *qua* changeable.[167] He rebuts the first objection which says that, if both the activities are in the changed and affected, the activity of each will not be in each but in another. He rebuts it as being able to be stated also against the opinion which says that the activity of both is one in the changed with regard to substrate. The rebuttal tries to show that the absurdity alleged against the hypothesis is not absurd. For it is not absurd for the actualization of one thing to be in another; rather it is necessary in the case of the changer and changed and generally the agent and the affected. For if teaching is in what is affected by the agent it is neither in the teacher nor cut off on its own from the learner, nor in the learner in such a way that it can be present in him without the teacher. Rather it comes about in the changed with the changer being present and active.[168] For as the activity of the moulder in moulding wax is, in the wax, the moulding itself through which the one moulds the other is moulded, and it is not possible to take the activities as two, so it is the case of teaching which is one with learning in substrate. Let us not be diverted by the condition acquired by change, such as of shape in wax and knowledge in the soul;[169] for these are not changes but the products of change. For being moulded and learning or being taught are one thing, and having been moulded or taught is another. And this product is in the changed and is present apart from the changer; it is analogous to the disposition and reason of the teacher who is himself also separate from the taught. However, the actualization through the teaching and being taught is one in the substrate, even if different in account, the one being taken as coming from the agent, the other as from the affected. For, being between both of them, change connects with the agent and the affected without being cut off. But if it be taken as of this alone or in that alone the mean will no longer be preserved. For that reason when we see it as cut off, regarding it as in the changed, since it subsists in it, we say that it is the actualization of the changeable *qua* changeable. But when we see it in its involvement, beginning with the changer, we say that it is the activity of the initiator of change in the changed, as it will be defined at the end of the account of change.[170]

A version of the text runs: 'it exists, however, also as not cut off, but of this in that'; Alexander says: 'Not simply and as cut off in something, but in something definite, and that in relation to which it is named; it is named in relation to the learning process.' It is surprising to me how he understood 'not cut off' as equivalent to 'not indefinite', in spite of having well expounded 'in something, however, and not cut off' as showing that it is not as separated from the learner nor as in the teacher alone, nor in isolation.[171] And in the case of this reading the addition should mean the same.

202b8 Nor does anything prevent the same actualization being one for two things, [not such that their essence is the same,] but as the potential stands towards what actualizes it.

He replied to those who say that it is absurd for there to be one change of both that there was one and the same actualization of the agent and the affected which differed in form. So he says that this also is not absurd, as it seems to be. For nothing prevents teaching and learning and, generally, agency and affection being a single actualization which is one in substrate, even if different in account. For whoever speaks of teaching speaks of nothing other than what comes about through the activities of teaching and learning, in which case it is impossible for the one to be and come about without the other, and he who speaks of learning says the same thing once again. So they are one in substrate even if different in account, just as the interval between two and four is one but their accounts are different, since that of four is being double in relation to two, that of two is being half in relation to four. Aristotle himself used a still more obvious example, saying that they were one in substrate, two in account,[172] in the way that potential is related to what actualizes it. By 'what actualizes it' he refers to what is completely actual. For that already is active as well. For if what it has become in actuality is what it was potentially and nothing else, and what it had been potentially is what it has become actually, both of these are one and the same in substrate, but the accounts are different. For the potential is that which can come to become what it is said to be potentially, but that is actual which is already that which it is said to be in actuality.

202b10 Nor is it necessary that he who is teaching should be learning, [even if the action and the affection are the same, but not the same in such a way that the account which states the essence is one, nor as in the case of a mantle and a cloak, but as are the road from Athens to Thebes and that from Thebes to Athens, as was also said earlier. For not all the same features are present in things that are in some way the same, but only in those whose essence is the same. So, indeed, this is not the case, even if teaching is the same as learning and to be learning the same as to be teaching, just as even if the interval between separated things is one it is not the case that to be at an interval from here to there and from there to here are one and the same.

In general neither is teaching the same as learning nor acting the same as being affected in a strict sense, but that which holds of them all – change.] For being the actualization of this in that and the actualization of that by this are different in account.

The second absurdity that he brought against those who say that the activity of the changer and the changed are one is that it follows that teaching is the same thing as learning; and on this an obvious absurdity seemed to follow that he who teaches, in virtue of this teaching is learning everything that he is teaching and that he who is learning is, in virtue of this learning, teaching everything that he is learning. The objections being such, he proves that all the consequences have no necessity. He begins with the last which says that if to act is the same as to be affected, or to teach the same as to learn, it is necessary that the teacher, being the same as the learner, should be learning as he teaches. He rebuts this objection by the fact that not all the same things follow for all those in which there is an identity, but only for those who share the same account. He includes examples of things which have the same account but different names, the mantle and the cloak; for these have the same definition, which is why all the same things follow for them; also examples of things which are the same in substrate but do not share the same account are the road from Thebes to Athens and that from Athens to Thebes. The way up and the way down are similar. For in these cases the interval and the position are one and the same, but the point of departure and the destination are different, so that their whole form is different, and therefore the same things do not follow. For the situation of one who ascends is one, of him who descends another. Also he who goes from Thebes to Athens and he who goes from Athens to Thebes encounter different things first and second. Therefore he who is going to Athens is not going to Thebes.

Nor does the same follow for those things that are only one in substrate, but only for those of which account and being and definition are the same. So even if to teach and to learn are the same thing only in substrate and not also in account it is not necessary for him who is teaching to learn whatever he teaches just because he is teaching. For that would follow only then if because to teach and to learn were one in account the teacher and the learner were the same in account. For then it would be necessary for him who was teaching to be learning since the account was one with respect to teaching and learning.

Then next he refutes the consequence alleged in the previously stated problem as not being necessary. It was that which said 'if teaching is the same as learning then to be teaching is the same as to be learning'.[173] He proves that this is not necessarily the case once again through a use of the same example. He takes extension, whether of places or of numbers, as analogous to teaching and learning, and to extend from here to there and thence to here as analogous to the acts of teaching and learning, as follows: as are teaching and learning, so is extension, and, as to teach and to learn, so to extend from Athens to Thebes or from Thebes to Athens, or from four to two or from two to four. So as in these cases extension is one in substrate but the accounts are different, and for that reason it is not necessary for extending to be the same, so in the case of teaching and learning, since they are the same in substrate and not in account, for that reason to teach and to learn are the same in substrate but not in account.

Alexander says: 'It is possible that "so indeed even if teaching is the same as learning" and what follows is said superfluously, as if he had said "and if it were granted that teaching is the same thing as learning it is not necessary that to be teaching and to be learning are the same; nor, if the extension between the separated things is one, is it already necessary that to extend from here to there and thence to here are one and the same thing".'

Then, recurring to the first consequence on which the rest depend, he refutes that also. It was this: if the activity of both is one, teaching is the same as learning. So he says that altogether not even the basis of the consequences is necessary. For, if the change is one, it is not necessary that at once the constituents in the change – teaching and learning or agency and affection – should be the same in the strict sense, because the relations of the agent to the result and the result to the agent are different in account. But how is the change one, and one in the strict sense, since it has often been said that it is one in substrate but not in account? Perhaps change is one not in substrate only but also in account, being in the changed as caused by the changer. Therefore the account is also one, being 'the actualization of the changeable *qua* changeable', and what causes the change and what it is in are not separated in it, but they constitute change both together. However, agency and affection are separated, one concerning the relation from the agent, the other that from the result. Alexander, however, explains this also in the same way as the previous ones. He says: 'For if the activity of both is one in regard to substrate it does not follow at once that teaching is the same thing as learning, but the substrate of both teaching and learning is the same; for one thing, the act of learning, comes about as one through both of them, but in fact teaching is not the same as learning.' He seems to understand 'of this in that and that through this' as applying to change. I, however, believe it to be said with reference to agency and affection, both this and 'different in account'. But what does he mean by saying that teaching and learning are even one in substrate? It is because the act of learning becomes one through both. For the act of teaching also comes about. In general all of the causes of one thing are not one in substrate, since if so then so are father and mother; but those things are one in substrate because the extension is one and the same from the agent in the affected.

202b23 So it has been said what change is, [both universally and in its varieties. For it is not obscure how each of its species will be defined. For alteration is the actualization of the alterable *qua* alterable. Still more plainly, it is the actualization of the potentially productive and passive, as such, both generally and again in variety, whether housebuilding or curing.] The same account will be given about each other sort of change.

He now sums up and reminds us of the arguments about change in the outline provided, according to which it was once said to be the actualization of the potential *qua* potential, at another time to be the actualization of the changeable *qua* changeable.[174] He says that this account fits not only change in general but also particular sorts. For it is not obscure, he says, that each of its species will be defined by this definition. He does not say 'species' as if change was being divided within a genus, for change is not a genus, but as being different particular varieties. He indicates also with one of the particular species, alteration, how one should fit the general account to the particular species and adds a further third definition of change,[174a] which he says is plainer by linking agency to affection. For he says that change is the actualization of the potentially productive and passive as such. For since it has been demonstrated that change is completed from the activity of both the changer and the changed he reasonably includes in its definition the activity of the initiator of change as well, not only that of the changed. But what does he mean by the activity of the potentially productive? For the potentially productive and initiator of change is not yet actually an initiator of change, whereas the changer is in actuality. Also in general the potential *qua* potentially active is rather changed but does not initiate change, if the definition of change has been well set out. So perhaps he stated the definition as of the potentially productive referring to the change communicated by the unchanging causes in relation to natural and artificial change. For all changers initiate change in so far as they are in actuality, being unchanged in that respect. But the cause which is particularly called unchanging is in all respects unchanging, since it is actual in all respects and exists eternally. But things that initiate change naturally and artificially do so in so far as they are in actuality in respect of their condition, being in that perfect. For it is necessary for the changer which is already a form to communicate it to the changed. They are themselves changed in so far as the previously dormant condition is aroused to the activity of initiating change, through its potential for activity, which all things have which are not at the same time transferring whole their own activity. This is present only in things that are eternally in actuality.[175] Therefore through the whole time of initiating change both potentiality and actuality are in them, and they are changed through their progress from being imperfect, but initiate change through the perfection of their condition and actuality. Nature, which is also so predisposed, gives their disposition from within in relation to each form of change, as Andronicus also said.[176] For even if water be heated by fire, still the natural constitution in the water first becomes warm and thus warms, or warms with itself, the substrate. So he said 'potentially initiating change' in the same way as he previously called that which is changed while its potentiality remains 'the potentially changed' and so similarly that which initiates change while its potentiality remains. It is also clear that according to this account it is true to say of things that thus initiate change that in so far as they initiate change they are not changed. For they initiate change in so far as they are perfect, but in so far as they are perfect they remain and are not changed. And again, they are changed in so far as they initiate change; for a nature which is increased will cause increase with respect to the same form, but a skill proceeds to activity from its dormant condition.

Having given his third definition he decides to sketch partial kinds of change also, which he loosely called particular, from this basis. For he does not mean the numerically single, but those which differ in form. But Alexander believes it to be a return to the more plain version of the account of alteration, offering what I think are implausible defences. For Aristotle manifestly is accepting it as the third definition of change, even if the text added in the book as from a marginal note seems to be muddled.[177] Having introduced house-building and curing as alterations he added that 'the same account will be given of the other kinds of change'. For Alexander well notes that nearly all artificial changes are mainly by alteration.

But the things that Aristotle says here universally about change being what they are, I think it will be clear from them to those who examine them that Aristotle did not disagree with his master <Plato> with regard to change in the same sense of the word. Rather, Aristotle is looking at change in natural transformation, and does not regard things that are constant as being in change, while Plato assigns change also to departure from being into activity, and does not hesitate to apply the term also to intellect, each appropriately to his own philosophy.[178] For Aristotle always deals with the plain and obvious, but Plato with things that are in any way of the sort that they are.[179]

Chapter 3 ends here in modern editions. The discussion of the nature of change is finished and the rest of the book is about the unlimited.

202b30 Natural science is about [magnitudes, change and time, and each of these must be either limited or unlimited, even if not everything is either unlimited or limited, such as an affection or a point; for perhaps there is no necessity for them to be either. So it would be fitting for the investigator of nature to inquire about the unlimited, whether it exists or not and,] if it exists, what it is.

The investigation is about everything that accompanies all natural things. Natural things are bodies and magnitudes changing in time, and it is necessary that a body, change and time, being magnitudes, should be either limited or unlimited. The aforementioned are also continuous and the unlimited is included in the continuous, since the divisible without limit is continuous. So it is reasonable that the investigator of nature should also say something about the unlimited as well.

It is clear that the investigation of nature is about bodies from the fact that all natural constructs are bodies; it is clear that it is also about change from the fact that these natural bodies are the things which have in themselves the principle of change. He himself makes this clear at the beginning of *De Caelo* where he writes: 'Just about the whole of the science of nature seems to be about bodies, magnitudes and the affections and changes of these.'[180] It is clear that it is also about time from the fact that all change is in time and time is the measure of change, as will be shown.[181] Having said that each of these must be either limited or unlimited he reasonably noted that this is not a contradictory division of entities or of all natural things, [divides the true and false in all cases][182] but the antithesis of limit to limitlessness is as is the antithesis of possession and privation, which does not divide the true and false in all cases, but in those of a nature to fall under it, and when they are of that nature. That it is not comprehensive he confirms from two examples, affection and point. For what he calls affective qualities in the *Categories*,[183] such as whiteness, blackness, heating, cooling and the like, are not as such either unlimited or limited, and this is clear from the fact that being limited and being unlimited are predicated of quantities, and these are not quantities. Nor is a point subsumed under the limited or unlimited, because it is not unlimited since it is indivisible, and it is not limited because it is itself the limit of a line. Perhaps the unit is similar, for it too is indivisible and the principle of number. He added 'perhaps' to 'there being no necessity for them to be in either', whether the unlimited or limited, because while affections such as whiteness or heat are in themselves neither unlimited nor limited (for these are not quantities as such, but it is quantities that are said to be unlimited or limited) they also incidentally become quantities insofar as their substrates are quantities. The natural scientist is concerned also with affections insofar as they are affections of bodies. That is why he says at the beginning of *De Caelo* (451,21) 'just about the whole of the science of nature seems to be about bodies, magnitudes and the affections and changes of these'.[184] So the natural scientist will also examine the limited and the unlimited in relation to affections, even if they are present in affections incidentally. So the investigator of nature ought to inquire about the unlimited, first whether the unlimited exists or not, and if it does, what it is.[185] For this is the correct order of the questions, as we were taught in the *Analytics*.[186]

202b36 A sign that inquiry into the unlimited is germane to this science is [that all those worthy of mention who have joined in this kind of philosophy have discussed the unlimited, and all treat it as some principle of existence. Some, like the Pythagoreans and Plato, treat it on its own, not as an attribute of something else, but as the unlimited being itself a substance. However the Pythagoreans place it in perceptible things, since they do not treat number as separate, and they say that what is outside the universe is unlimited; but Plato says that there is no body outside, nor ideas, since these are not anywhere, but that the unlimited is in both perceptible things and] the latter.

After the evidence of the facts he confirms, as was his custom, that inquiry into the unlimited is germane to the science of nature from the opinion of the more distinguished naturalists. He shows that not just anybody but those worthy of mention who have joined in the study of nature have investigated the unlimited, not just as a side issue, but have all treated the unlimited as some principle of existence. So it is necessary for those studying the principles of nature to include a discussion of the unlimited. Next he sets out the different opinions which his predecessors held about the unlimited, and also what was common to them. He states the first difference through which some supposed the unlimited to be some substance on its own and not an attribute of something else, as did the Pythagoreans and Plato, who were agreed on this matter.[187] But most of the naturalists treated being unlimited as an attribute of something; some supposed air to be unlimited, like Anaximenes and Diogenes,[188] some water, like Thales,[189] some the intermediate stuff, like Anaximander.[190]

453,1 These and those like them treated being unlimited as an attribute of magnitude; but the followers of Anaxagoras and Democritus say that the unlimited arose in the number of principles, the former of ones like what they composed, the homoiomeries,[191] the latter of atoms.[192] The only difference is that these too are obliged to say that the unlimited is also in magnitude. He says that Plato and the Pythagor-
5 eans differ insofar as the Pythagoreans placed being unlimited in perceptible things and not as separate. For they say that the unlimited is also in numbers, but that number and mathematical objects in general are thought of on their own but do not exist on their own, but exist in perceptible things; and they say that perhaps that number in which they see limitlessness or process without limit is in percep-
10 tible things. This, of course, is not that number which they hymn, saying 'Listen glorious number, father of the blessed ones, father of men',[193] nor the one which Hippasus distinguished as being the first paradigm for the making of the cosmos.[194] He says that they held this unlimited also to be outside the heavens and clearly not perceptible,
15 since Timaeus himself says[195] that no body can be thought of as outside the cosmos. So they seem to say that both number and the unlimited in number, which is what they said was the even, were double, the one perceptible, the other intelligible, which they said was outside the heavens.

However, Plato places no body outside the heavens. He says clearly
20 in the *Timaeus* of the cosmos, which the heavens contain from outside: 'He who constructed it constructed it out of all fire, water, air and earth, leaving no part or power of any of them outside.'[196] He says that the forms also are not outside the heavens 'because they are not anywhere' nor in any place whatsoever. However, he says that limitlessness is in both perceptible things and the forms. For they say that
25 Plato states that the one and the indefinite dyad are the principles of perceptible things, but he placed the indefinite dyad also among things intelligible and said that it was unlimited.[197] Also he posited the great and the small as principles and said that it was unlimited in his lecture *On the Good* at which Aristotle, Heracleides, Hestiaeus and others of Plato's circle were present.[198] It was said that they wrote
30 up what was said enigmatically; but Porphyry, who promises to set it out clearly, wrote the following about it in his *Philebus*: 'He himself assigns the more and the less and the intense and the gentle to the nature of the infinite. For where these are present, proceeding by increased and decreased tension, they neither are static nor limit what partakes of them, but they proceed into indefinite limitlessness.
35 It is the same with the greater and the lesser and the great and the small, which are the terms which Plato used as equivalent to them. For suppose we take some limited magnitude, such as a cubit, and
454,1 cut it in two; if we were to leave the one half-cubit uncut but cut the

other half-cubit in short pieces and added them to the uncut one, there
would come to be two parts of the cubit, the one proceeding to being
lesser, the other to being greater, without end. For we should not come
to an indivisible piece as we cut; for the cubit is continuous. But the
continuous divides into ever divisible parts. Such an unending cut 5
makes clear that some natural limitlessness is enclosed in the cubit,
or, rather, more than one, one proceeding to the great, the other to
the small. In these the indefinite dyad is also seen, consisting of the
unit tending towards the great and that tending towards the small.
These features are present in both continuous bodies and numbers.
For the dyad is the first even number, and the double and the half 10
are both contained within the nature of the even, but the double in
excess the half in deficiency. So excess and deficiency are in the even.
But the dyad is the first even number and is as such indefinite, but
is made definite by participation in the unit. For the dyad is definite
insofar as it is one single form. So the unit and the dyad are the 15
elements of number, the former limiting and informing it, the other
indefinite and involved in excess and deficiency.'[199] That, in almost
his own words, is what Porphyry said, having promised to make
distinct what was enigmatically said in the conference about the good,
also perhaps because that was in agreement with what is written in
the *Philebus*.[200]

Also Alexander himself, acknowledging that he is speaking on the 20
basis of Plato's lectures on the Good which were reported by Aristotle
and others in Plato's circle, wrote as follows:[201] 'When Plato was
seeking for the principles of things, number seemed to him to be
primary in the nature of all other things; for the limits of a line are
points, while points are units having a position; but without the line
there would be neither surface nor solid,[202] while number can exist 25
without them. Therefore, since number is primary in the nature of all
other things, he thought this to be a principle, and that the principles
of the first number were the principles of all number. But the first
number is the dyad[203] of which he said that the principles were the
unit and the great and the small. For, insofar as being the dyad, it
contained within itself both multiplicity and smallness. Insofar as it 30
exhibits doubleness it contains multiplicity; for doubleness is multiplicity and excess and a magnitude; insofar as it exhibits being half
it contains smallness. Therefore in these ways there are excess and
deficiency, and the great and the small are present in it. But insofar
as each of its parts is a unit and it is itself a single form, the dyadic,
it partakes in the unit. That is why he said that the principles of the
dyad were the unit and the great and the small. He called it the 35
indefinite dyad because as sharing in the great and the small or the
greater and the smaller it contains both the more and the less. For
these proceed through increase and relaxation of tension and are not 455,1

stable, but continue towards indefinite limitlessness. So, since the dyad is the first of numbers and its principles are the one and the great and the small, it is necessary that these should also be the principles of all numbers. But numbers are the elements of all that there is. So the one and the great and small, or the indefinite dyad, are also the principles of all things. For, indeed, each number, insofar as it is a particular thing, and one, and definite, partakes of the one; insofar as it is divided and a multiplicity it partakes of the indefinite dyad. Plato also called the ideas numbers; so he reasonably made the principles of numbers also the principles of ideas.[204] He said that the dyad was unlimited by nature because the great and the small, or greater and lesser, have no bounds but contain the more and less, which proceed without limit.

Thus Aristotle has said in what things the Pythagoreans, and in what things Plato, placed limitlessness; next he investigates what the Pythagoreans said that the unlimited was, and why they did so.

203a10 The Pythagoreans said that the unlimited was the even; [for this, they say, being isolated and limited by the odd, supplies things with their limitlessness. A sign of this is what happens in the case of numbers. For, if gnomons are placed round the one and separately,[205] in one case the form becomes different on each occasion, in the other the same.] But Plato said that there were two unlimiteds, the great and the small.

He has said in what things the Pythagoreans, and in what things Plato, placed limitlessness, that the former place it in the perceptible and outside the heavens, but he in perceptible things and in ideas. Now he adds what each of them says the unlimited is, starting with the Pythagoreans. These said that the unlimited was the even number, because, according to their expositors, everything even can be divided into equals, and what is divided into equal parts is unlimited through bisection. For division into equal halves proceeds without limit. But when the odd is added it limits it; for it prevents its division into equal parts. So in this way the expositors ascribe limitlessness to the even through the division into equal parts, and it is clear that they take the unlimited division not as being of numbers but of magnitudes. For numbers, including the even, do not all divide into equals for long, and those that are divided reach[206] the one and halt the division, while in the case of magnitudes what prevents what is left from being continually divided, even if the division is not into equal parts? For e.g. the third part of any given can be again divided into a double part and a third. This can go on without limit, both by division and by addition of the divisa. In general Aristotle does not seem to ascribe being unlimited to the division into equal parts.

Perhaps in every cut the even is the cause of every division; for every single cut divides what is cut into parts, equal or unequal. So since 456,1
the cut into two, carried on continually, produces a progression without limit, and the dyad is the principle of the even, for that reason they said that being unlimited was through the even. Similarly the even is the cause of addition; for one is added to one. For this reason Aristotle as well says that the even 'being isolated and limited by the 5
odd, supplies things with their limitlessness'. For each body contains its being divisible without limit through the even which is contained and, as it were, hidden within itself. For it contains the cuts without limit potentially and not in actuality. That is why he says that the even is contained and supplies things with their limitlessness. For to 10
the extent that each of the divisa is in some things, and each of them itself is always limited as it is by the odd, and each is in the whole of the bodily form of things perceptible, it is divisible through its matter and its substrate and is in this way even and unlimited, but in that it is a definite particular it is and undivided, and in that way odd. 15

As a sign of the odd number supplying form and providing a limit and of the even being material and unlimited in kind, the Pythagoreans take what happens by the addition of numbers. For as odd numbers are added in turn to a square number they preserve its squareness and symmetry, so that it is increased only in quantity. But when an even number is added it alters the form for the square 20
number, making it oblong, being made greater on a different side at different times. For let the numbers be set out in order from the unit to ten: 1, 2, 3, 4, 5, 6, 7, 8, 9, 10. Of these it is clear that one is the first odd number and the first square, since one times one is one. But of the numbers set out the odd after one are 3, 5, 7, and 9, the even are 2, 4, 6, 8, 10. So if we add 3 to one, which is the first square we get a 25
square 4 with sides equal to two. And again, if we add the next odd number 5 to the square 4 we get 9 which is itself a square. And if we add 7 to this we get the next square, 16, and adding 9 to this we get 25. All are successively square and equal all round. But if we add 2 30
to the square unit we get 3 which is no longer a square, and if to the four-unit square we add 4 or 6 we get 8 or 10 which are oblong. For the sides are twice 4 or twice 5. Also it is clear that to keep the same form and to produce symmetry results from limit, but not to keep the 35
same form and the even numbers that increase the form in different ways by amounts that make an oblong stems from limitlessness.

The Pythagoreans called the odd numbers gnomons because when 457,1
placed around squares they preserve the same shape, like the gnomons in geometry.[207] For they call the two complements together with one of the parallelograms around the same diagonal a gnomon, which, when added to the other of the parallelograms around the same diagonal makes the whole similar to that to which it was added. So 5

the odd numbers are also called gnomons because, when added around the already existing squares, they always preserve the square shape. So when the gnomons, i.e. the units according to the succession of odd numbers,[208] (for it is these that are gnomons) are placed around one square number, sometimes separately, sometimes the even numbers separately, when the latter are added the form continually becomes different from the original; but when the gnomons are added it is one and the same as the former. But Alexander well added to the explanation, that 'when the gnomons are put around them' indicates making diagrams using odd numbers, while 'and separately' indicates arithmetical addition without diagrammatic placing around in the case of even numbers. For the Pythagoreans are accustomed to using diagrams. With regard to the surrounding of monads, they write down the numeral 1 and add to it three 1s in the form of a gnomon, which results in a 4 drawn in a square like this:

1	1
1	1

Then again they put five 1s round these as equivalent to being five monads larger and make a square with three to a side:

1	1	1
1	1	1
1	1	1

But putting even numbers around does not produce a definite figure. So he does not propose to put the odd numbers around like the even ones in the drawing, but to add them arithmetically. So that 'and separately' is understood as equivalent to 'and without putting them around' by supplement in the case of even numbers. But perhaps it is possible to understand in both cases[209] both 'with gnomons put around' and 'separately' as referring to both even and odd numbers, so that he will be saying that both with the gnomonic enclosures and also without in arithmetical combinations of both even and odd numbers it is possible to see the odd ones preserving similar forms but the even ones making them dissimilar. For even if even numbers are not strictly called gnomons since they do not preserve the same form, but still, if put around like gnomons, they show the dissimilarity even in diagrams. He said 'putting gnomons around the unit' either because the first enclosure is around the monad or because the putting around of more monads is always about a single number.[210] Having said what the Pythagoreans said that the unlimited was, that it was the even, he added what Plato said it was, that it was the great and the small or the greater and the lesser; for each of them proceeds without limit. But he did not take each of them separately, but took great and small together as the same thing, which he called the indefinite dyad, whether he called matter or the column of the unlimited by that name. That is why he placed the dyad also among the ideas, since they were also numbers.[211]

203a16 The students of nature all suppose some nature of the unlimited other than [that of the so-called elements, such as water or air or what is between them. Of those who make the elements limited in kind none make them unlimited in number. Those who make the elements unlimited in kind, like Anaxagoras and Democritus, the former homoeomeries, the latter from the supply of all possible shapes,] say that the unlimited is continuous by contact.

Aristotle was accustomed to calling those who were engaged in this portion of philosophy, the natural, naturalists (*phusikoi*), and of these especially those who used, especially or alone, maternal principle. So these naturalists who made matter the basis of things that come to be, and regarded the unlimited as surrounding it, as one would expect did not also say that the unlimited was a substance, but an attribute. Of these some supposed a single element and said that it was unlimited in magnitude, as did Thales of water, Anaximenes and Diogenes of air, and Anaximander of the intermediate;[212] others made them unlimited in quantity like Anaxagoras and Democritus and introduced the unlimited both in quantity itself and also in magnitude.[213] For the unlimited in quantity were held together by contact, but not by unification, and constitute an unlimited magnitude. Those who made them [the principles] more than one in number, but say that the principles are limited in number, like Empedocles, these introduce the unlimited neither in regard to magnitude, nor in one nor in many nor in their combination. For if they are limited in number and each is limited in size, whence should they gain limitlessness? He says that the so-called elements are fire, air, water and earth.[214] Having mentioned water or air as being among the much discussed elements, he added 'or some intermediate between these', water and air, which Anaximander supposed. But perhaps he said 'the so-called elements' not about the four now accepted but about those supposed by the naturalists, whether water or air or some intermediate between these, which they supposed to be an element in things that came to be. Also Alexander explains 'so-called' in another way as equivalent to 'those things which merely were said and seemed to be elements but were not', because these are compounds and come from elements, matter and form.

He did not say 'all students of nature' simply, but of those who say that there is some unlimited, since he knows that those who make the elements limited are also called naturalists, like Empedocles. But perhaps, as was said earlier, he thinks that those are truly naturalists who are concerned with the material cause, among whom one would not think it right to include Empedocles, who introduces strife and love as efficient causes. Perhaps also not Anaxagoras, who says that intellect distinguishes the homoeomeries, except that at least in the statements of the causes he seems himself not to use intellect as well, but more material ones. But he says of those who make the elements unlimited in quantity, such as Anaxagoras who supposed the homoeomeries to be unlimited in quantity and Democritus the atoms, that they manifestly introduce limitlessness in quantity, which Aristotle indicated by saying 'those who make the elements unlimited'. That they do so also in size he shows by saying that they say that the unlimited is continuous by contact and not by unification, as do those who make the element one and unlimited. It becomes clear that the continuous by contact is something unlimited from the following; for things unlimited in quantity that have a magnitude and are of the same form, such that they also touch each other, make a magnitude continuous by contact. That is why Eudemus, in the second book of his *Physics*, says: 'To say that what is of the same form is unlimited in quantity is in no way different from their being so in magnitude.'[215] Aristotle said of the atoms of Democritus that they were a universal seed-bed of shapes,[216] since he supposed that the differences of atoms in shape also were unlimited.

203a23 And Anaxagoras said that any part [was a mixture like to the universe because he saw anything as coming from anything. That, it seems, is why he says that all things, such as this flesh and this bone, and so on, were once all together. Therefore all things also and all at once. For there is a beginning to the distinction not only of each individual thing but of them all. For since everything that comes to be comes to be from some body of the same sort, and everything comes to be, though not all at once, there must also be some principle of coming to be, and this is one, which he calls intellect. But intellect works by intellection from some beginning; so that it is necessary that things were once all together and] things started to change at some time. 30

Since some of the naturalists held that the principle was one, others that there were many, but limited in number, others that they were unlimited in number, he said that those who held that it was one said that the unlimited was an attribute in magnitude of it. But, they said, those who posited that the elements were limited in both number and magnitude denied that there was an unlimited, whether 35 in number or in magnitude. So next he says of those who make the elements unlimited in number that they do not introduce only the 460,1 unlimited in quantity, but also that in magnitude. For the elements unlimited in quantity are magnitudes in actuality and held together by contact, and so make the magnitude unlimited as was said earlier.[217] But since each of Anaxagoras and Democritus supposes as principles unlimited in quantity, the one homoeomeries, the other 5 atoms, he first gives an account of the opinion of Anaxagoras, and tells us the explanation why Anaxagoras came to such an idea, and shows that it is necessary that he should say that not only the universe should be a mixture unlimited in magnitude but also each homoeomery which has, like the universe, all the parts within it, and not only things without limit, but also unlimiteds without limit.[218] 10 But Anaxagoras came to such an idea because he believed that nothing came from what is not and that everything was nourished by its like.[219] So he saw that everything comes from everything, if not immediately then still in order (for air comes from fire and water from air and earth from water and stone from earth and again fire from stone), and that when the same nourishment is provided, like bread, 15 many different things come to be, flesh, bones, sinews, veins, hair, talon and perhaps feathers and horns, but like is increased by like. For these reasons he supposed that these things were in the food and in the water if trees, wood, bark, leaves and fruit were nourished by it. Therefore he said that everything was mixed in everything, and that things came to be by being separated out.[220] Perhaps he included 20 in addition also that while things continued other things came from them, like fire from stone and air from bubbling water. So he saw that all things were separated out from each of the things now distinguished from each other, like flesh, bone, etc. from bread, since all things were present in it at once, mixed together, and he got from these the idea that all things were mixed together before they were 25 distinguished. That is why he began his book with 'Everything was together',[221] so that anything whatsoever, such as this loaf of bread was a mixture of this flesh and this bone like the universe. For he came to the mixture of all from the mixture of each. For particulars are more obvious and more familiar to perception than are wholes.[222] Further, he observed the likeness of each thing to the whole in the case of this mixture and their distinction, both causal and temporal, as coming from a beginning. So just as in each thing there is at some time a principle to distinguish it, so also in the universe. And just as from each something appears to come to be at some time, so, since everything comes to be, the coming to be of all things was separated out from the universal mixture, even if not all together. But since bodies are externally changed and are in a way on the same footing they needed some principle and cause superior to them, which Anaxagoras calls intellect.[223] For that which was distinguishing the complex and bringing out in order the distinction from the complexity had first to contain itself the existent as a complex and reveal the distinction in itself. Such was intellect, which he thought of as working from some principle in matters in which he says the intellect commenced the distinction.[224] The idea through which Anaxagoras came to this hypothesis about principles thus became clear. From what has been said it is easy to understand that if everything is separated out from everything, and everything is in everything, not only universally but in each particular, there will be unlimiteds without limit, not in number only but also in magnitude.

Aristotle gives an account of the opinion of Anaxagoras as it is plain to see; but Anaxagoras, being a wise man, gave a riddled double account of the imposition of order, the one unified, intelligible and pre-existing, not in time, (for this one is not temporal), but in superiority of being and potentiality the other distinct from this; and in accordance with this it comes to be through the demiurgic intellect. This was said in the comments on Book I,[225] where I tried to set out the evidence from the statements of Anaxagoras himself. Reasonably, each perceptible thing shares in all things and participates in them. 'For there is one conflux, one joint breath, all things feel together',[226] as Hippocrates says, because of the original unification existing before them in the intelligible realm. For when distinguished these things are not wholly torn apart from each other; for distinction is not total sundering. Therefore walking, or colour, or affections in general, and dispositions cannot be separated from their substrates. But the appearance of saying that the beginning of distinction was temporal was characteristic of the ancient naturalists and theologians, as they conformed to the weakness of our intellect. For we are not capable of extending our intellect to the extension of eternity, but we require some beginning to be posited and then to see the consequences in turn.

203a33 But Democritus says that none of the primitive things come one from another. [But the common body is the principle of all things alike, differing in its parts in size and shape.] From these considerations it is clear that this study is appropriate for naturalists.

He has told how the positing of a principle unlimited in magnitude also followed for Anaxagoras from his saying that there was an unlimited quantity of bodily principles, the homoeomeries, and now he shows that, even though he did not make coming to be a separating out, nor said that everything was present in everything like Anaxagoras, still Democritus also, having posited as principles atoms unlimited in quantity and the void, himself no less than Anaxagoras said that the principle was unlimited not only in quantity but also in magnitude by contact, and that according to him the unlimited so becomes contiguous. For the things that are unlimited in number have some magnitude and are homogeneous, so that they also are in contact with each other, and so make an unlimited magnitude, as has been said often. Aristotle showed that the atoms have a magnitude according to Democritus, from his saying that they differ in shape and magnitude. How would something be different in magnitude if it had none? And how in shape if it were not given a shape by lines or surfaces? It is utterly clear that what is contained by lines or surfaces is a magnitude. That the things unlimited in quantity are in contact followed immediately for Anaxagoras since, as he said, 'all things were together'.[227] It follows for Democritus insofar as the common body of atoms was one for all, as he put their difference in size and shape and not in a substrate body. From this it follows that one should say that they are of the same kind and, through that, are in contact. For things of the same kind move to the same place if not prevented and touch each other,[228] while the void is not a barrier to contact. For then the void would be a greater unlimited than the unlimited body, whether with it or without it, whence it would be a barrier. But there is not a greater unlimited than the unlimited. Also the void has absolutely no power to be a barrier.

So, therefore, if virtually all the naturalists accepted an unlimited magnitude which also was of the character of a principle, as being at the beginning, it is clear that an account of the unlimited would be appropriate for one who is discussing the principles of nature.

203b4 They reasonably all also treat it as a principle; [for it cannot be pointless, nor can it have any other power save as a principle. For everything is either a principle or derived from a principle, but there is no principle of the unlimited] for it would be a limit of it.

Having said that the ancient naturalists in introducing the unlimited introduced it as a principle, he now proves, not that there is the unlimited, but that those who supposed it all treated it reasonably as having the character of a principle. He proves it by a division; for everything whatsoever is either incidental and pointless, like the automatic, or it is as such; and if it is as such it is either a principle or derived from a principle; so if the unlimited is neither pointless nor derived from a principle it is clear that it is a principle. It is clear that it is neither pointless nor automatic, since everything such is rare, but the unlimited is everywhere and everything and not rare. Also, if the unlimited is also everlasting, as will be proved from its not coming to be, it could not be pointless or automatic; for nothing such is everlasting. But certainly the unlimited is not derived from a principle nor is there a principle of the unlimited, for it would be a limit of it, as he says, and so it would no longer be unlimited. People think that this argument contains a fallacy, since they take Aristotle to say that if the unlimited had a principle (*arkhê*), and a beginning (*arkhê*) is with regard to magnitude and a limit, it would have a limit; so that it will not be unlimited but limited. They say that the mistake arises from the ambiguity of '*arkhê*'. For as the distinction was made in the first book[229] *arkhê* means both the beginning of some thing, as they now took it, but it also signifies a causal relationship. For matter, form, agent and end are all *arkhai*.[230] So when he had set out to show that the unlimited has not got an *arkhê* which is a causal principle, but that it is itself a principle, the *arkhê* as causal principle was exchanged for that as the beginning of a thing, in which case it is true that if it has an *arkhê* it also has a limit. For it would no longer be true in the causal case that, if something has a principle in the causal sense, it has a limit as an existing thing. For that limit does not follow from that sort of principle. But to say that he is trying to support this opinion persuasively seems to me to be in fact unpersuasive. For he seems to be saying that it is a necessary consequence that if the unlimited has a principle it also has a limit.

Those matters are better stated by Alexander, to the effect that if the unlimited is supposed to come about from an *arkhê* of whatever sort it certainly has a limit in magnitude, so as to be no longer unlimited. For if it came from a temporal beginning (*arkhê*) it would not come to be as a being instantaneously, but from some beginning in magnitude, as animals, plants and all products are seen to come about. But if it has a beginning in magnitude it would also have this as a limit. Even if it had a beginning which was like an element and were to come from this beginning, as from matter, then if the unlimited had a limited beginning within itself, its matter, it would also itself be limited; but if it is unlimited the unlimited will be a principle. Similarly if the unlimited were to have a beginning with regard to form. For if form is the principle of the compound, then, if the form is unlimited, the unlimited is a principle, but, if it is limited the unlimited would be in the compound in respect to matter. For matter is also a principle. So the unlimited is a principle in that case also. For the compound, of which form is a principle, is nothing other except matter and form. And, even if the unlimited were to have a creative principle, it would invariably be a body, because it is also the principle of the thing from whence it began to come to be,[231] which would also be a limit. Also in any case it would be something that had come to be and would have a limit at which its coming to be ended. But if the unlimited magnitude had a principle which is an end and something for the sake of which, then, first, nothing prevents it from being itself a principle if it has such a principle. For also the efficient cause and the elements have a final principle. For Plato also when seeking the final principle of the creation of the universe finds it in the goodness of the creator where he says 'let us say for whatever reason he who composed it composed coming to be and this universe; he was good'.[232] Then how should the unlimited magnitude, which is indeterminate, be referred to some determinate end? But if as a magnitude it were to have a determinate end it would have itself to be determinate as a magnitude. But the boundary of a magnitude is its circumscription and its shape. It is clear that Aristotle draws the conclusion from the magnitude's being unlimited that it does not have a principle, not merely as an element, but also neither a creative nor a final one. This is clear from what he says of those who do not posit other causes beyond the unlimited, such as intellect or strife and friendship or anything of that sort, that they say that the unlimited 'contains all and governs all',[233] so that, if it is unlimited, it does not have a creative or final cause. So his saying has become clear, that if there were a principle of any sort of the unlimited magnitude it would be a limit of it as a magnitude.

203b7 Also as being a principle it is uncreated and indestructible. [For what has come to be must have some end, and there is a terminus to all decay. Therefore, as we say, it has no principle, but it seems to be one of other things and to contain all and govern all, as they say who do not suppose other principles beyond the unlimited, such as intellect and friendship; and this is the divine, for it is deathless and indestructible,] as Anaximander and most of the naturalists say.

He has said that those who posit the unlimited reasonably say that it is a principle; for either it is pointless, or derived from a principle or is a principle, and when the others have been refuted it remains that it is a principle. Now he adds something further of what is said to be present in the unlimited, that the unlimited should be called uncreated and indestructible, if it exists at all. This also supports its not being pointless, since the uncreated and indestructible is everlasting, but the pointless is not everlasting, being incidental and rare. The Socrates in the *Phaedo* clearly taught that it is fitting for a principle to be uncreated and indestructible. 'For', he says,[234] 'if a principle were to come to be from something it would not still be a principle'[235] and that 'if a principle has been destroyed it will never come to be from anything else nor will something come to be from it'. That being a principle belongs also to the uncreated is clear. For if it is not a principle it comes from a principle. He proves that the unlimited, if it exists, is uncreated and indestructible as follows: if it has come to be its coming to be must have been completed at some time, or it would not have come to be. So if its coming to be is not instantaneous it is clear that there is also some beginning of its magnitude, from which it begins to come to be, and an end to it, at which its coming to be terminates; and when it is completed it will no longer be coming to be, but will have come to be. Also if its magnitude were at some time to be being destroyed it will also at some time have been destroyed. So there will be some limit to it, and when the extremity of it has been destroyed the whole of it will have been destroyed. So if there is an unlimited it is necessary for it to be uncreated and indestructible; this is peculiar to principles and not to what is derived from a principle. So those who posit the unlimited alone as a principle and do not argue for some other causes, as Anaxagoras did for intellect and Empedocles for love and strife,[236] were satisfied with the nature of the unlimited and this material principle for the coming to be of everything, on the ground that from its inexhaustible supply there will always be coming to be and everything will be contained and governed by it. For if this alone is posited as a principle it serves the purpose of a principle that is not only material but also efficient and final, which Anaxagoras attributes to intellect, Empedocles to love and strife and to necessity.[237] Also they say that the cause is divine both as a principle and as uncreated and indestructible. As such Anaximander posited the unlimited between fire and air as a principle.[238] It is nothing strange if he called it divine, but rather even necessary. For from this he showed that god was superior to it; for the divine is what partakes of god. The discussion among such men was about natural principles, as for Aristotle in this work, and not about those superior to nature. It is nothing surprising if they said that this contained and governed; for it belongs to the material cause to contain as it extends through all things, and to govern the things that come from it as through its suitability.[239]

203b15 The ground for there being an unlimited [can be viewed from five aspects; from time, since this is unlimited; from division of magnitudes, since mathematicians make use of the unlimited; from the fact that only if that from which coming to be is subtracted were unlimited would coming and ceasing to be inexhaustible; also since the finite always has a limit at something, so there must be no final limit if it is necessary that anything should always be limited by something else. But most of all and most importantly is what provides a problem common to all; for because it is never exhausted in thought, number seems to be unlimited as also mathematical magnitudes and that without the heavens. But if what is outside is unlimited body also and universes seem to be so as well. For why rather here in the void than there? So if there is a mass in one place it will also be everywhere. At the same time, if there is also a void and unlimited place, body must be also;] for there is no difference between possibility and existence among things everlasting.

He has shown that an account of the unlimited is necessary for the natural scientist, both because of the concept of the unlimited itself and because of the opinion of the earlier naturalists; so next he turns to an inquiry about it, preserving the order of the problems and first inquiring whether there is an unlimited or not. In addition he thought it useful to trace out the notions from which some came to posit the unlimited so that if they should appear to have a bearing on some[240] matter we also should accept them. But if they should be caught stepping on emptiness and following the indefinite content of the imagination, we should refute them and show that the unlimited is a figment. He says that there are some five grounds on which we believe in the unlimited, not only that in magnitude but an unlimited of any sort. For having begun to trace completely the notions from which we supposed the unlimited he lays them all bare, not only those with regard to bodily magnitude, but also those regarding any magnitude, and those which fashion, sometimes truly sometimes falsely, a numerical unlimited. He sets out the first notion of the unlimited as that which is derived from the actual limitlessness of time. For if time is not unlimited, once there was no time and there will be when there is none. But the past and future of time will once again be parts of it, so that there is time also when there is not; for there will always be time. So time is unlimited.

The second confirmation of the existence of the unlimited is from divisibility of magnitudes, or, as Eudemus says, of continuous things.[241] For it seems to be inexhaustible. But it is the unlimited that is inexhaustible. Also the mathematicians make use of the division of magnitudes, sometimes treating it as an axiom; for they divide every magnitude and every line in any ratio whatsoever, taking the given line of any length, on the ground that division of magnitudes was unlimited. But sometimes also they prove it. For if, as they say, a straight line inscribed in a circle never becomes equal[242] to the circumference, it is clear that the area cut off between the straight line and the circumference is divisible without limit. For if the division comes to an end the straight line will coincide with the circumference. Also if two sides of a triangle are always greater than that remaining, then in a right-angled isosceles triangle in which the base is greater than either of the other sides, if we take off an amount equal to the base from both sides and draw a line parallel to the base through the resultant points we shall make a right-angled isosceles triangle similar to the whole. And we shall be able to do this for ever. For we shall never cease from making the triangle smaller nor from finding the two sides greater than that remaining one, so that we can take from both an amount equal to the base and draw a line parallel to the base through the points. But if it is possible to do this without limit the area of the triangle is cut without limit. But if there is division without limit there is an unlimited in magnitude. For it would not be divided without limit unless there were an unlimited.[243]

He says that the third of the considerations in favour of the unlimited is that only thus would coming to be not be exhausted, if that were unlimited by subtraction from which what comes to be comes to be. For this persuaded some of the naturalists as well to suppose that the element was unlimited.

The fourth, which is hard to deal with, is that everything limited appears to reach a limit at something. For if everything limited reaches a limit at something else outside it, that is either limited or unlimited. If it is unlimited we have it from this very fact that the unlimited exists; but if it is limited like the earth, then this reaches a limit at something, and so without limit. And if it does so without limit, then there is the unlimited. For no final limit will be found, since that too will be limited at something else. As Alexander says, 'it is especially on this argument that the followers of Epicurus relied in saying that the universe was unlimited, because everything limited has something outside it that limits it'.[244] Aristotle refers to the argument as being very ancient.

Aristotle said that the fifth is the strongest and most important and provides a problem common to all. The power of our thought or imagination can always add something and take away something and is never defeated and exhausted. Number seems to increase without limit because of this fact that we can add a unit or a number to every given number, and that mathematical magnitudes can divide and increase without limit, because every one taken can be cut and the segment can be added conceptually, though not, indeed, in practice. And, since concepts seem to come to be with relation to things, we think that things are as we imagine them. We are similarly affected also about what is beyond the heavens; for we always conceive of something beyond what is given, so that what is beyond seems to be unlimited. Also, if it is void, as Democritus seems to have said, there would be an unlimited number of universes as well;[245] for why in the void rather here than there? So if in one place the mass of body would also be in all places, so that it too would be unlimited. And in any case, if the void is a place capable of receiving a body, in the case of the everlasting what can come about would unfailingly come about, so that if it did not come about it would also not be possible; for, in the case of perishable things, nothing prevents them perishing before what is possible reaches reality, but in the case of the everlasting this is impossible. Therefore there is no such thing as possibility among everlasting things, but only among things that come and cease to be, so that if the void is unlimited there will also be unlimited place and unlimited body, which unlimited place is of a nature to receive.

'But Archytas', according to Eudemus, 'raises the question as follows: "if I have reached an extremity, such as the fixed heavens, can I stretch out my hand or a staff outside it, or not? It is absurd that I could not stretch it out; but, if I stretch it out, what is outside will be either body or place. It makes no difference, as we shall learn. So he will always continue walking in the same way to the limit chosen on each occasion, and ask the same question, and if there will always be somewhere else to which the staff reaches, it is clear that it is also unlimited. If it is a body, the thesis is proved; if it is a place, and place is that in which a body is or could be, but what is potential must be treated as actual in the case of things everlasting, then thus also there should be unlimited body and place".[246] Perhaps this argument is a grim problem for us as well, who say that there is nothing outside the heavens, on the ground that the cosmic body, whose limit is the heavens, occupies the whole[247] of space. So if, having reached the expanse of the heavens, he were to stretch out his hand, where would it be stretched out? Surely not into nothing, for no existing thing is in what does not exist. But nor will it be prevented from stretching out; for it also cannot be prevented by what is nothing.

So Aristotle in this way says that the confirmation of the existence of the unlimited depends mainly on five reasons, but Eudemus says that it is on six.[248] Having said that the first is division without limit of the continuous, he says that the confirmation seen through addition of numbers is the counterpart to it. But Aristotle did not omit this explanation either, but put it aside among unlimited things believed in through conception, when he said that 'number also seems to be unlimited', and it is clear that mathematical magnitudes can be divided without limit. But one might raise the problem why he says that the addition of numbers and the subtraction of magnitude without limit are conceptual. For from conception the truth about things appears. For to what number cannot one add and what magnitude is not divisible unless magnitude consisted of points? Or is it not process without limit but an unlimited which he says is through conception, because of the conception of increase and decrease without limit? For process without limit exists in both directions, but not yet limitlessness itself, as will be proved. For neither number nor magnitude is unlimited.

203b30 But the study of the unlimited presents a problem. [For many impossibilities arise both for those who hold that it exists and for those who hold that it does not. Also in what way does it exist; as a substance or as an essential attribute to something natural? Or is it in neither way, but none the less there is an unlimited] and[249] things unlimited in number?

He has said from how many and from which considerations those who supposed an unlimited did so, and he intends soon to show in what way the unlimited exists and in what way it does not. In between he says that on both sides problems arise, for those who say that the unlimited does not exist and for those who say that it does. Also, if it is something for those who treat it as a substance and those who treat it as an attribute, or for those who say it is neither of these, they nonetheless say that there is the unlimited in both number and magnitude. For this is demanded by the inexhaustibility of division, and is particularly characteristic of process without limit. So what problems does he show to face those who say that there is no unlimited? Is it necessary to abolish the unlimited division of continuous things? But, if this is abolished, geometry, which uses it as a principle, is abolished, and the view is introduced that magnitudes consist of atoms and points and time of instants, which will be refuted in many ways. Also the unlimited increase of numbers, which is plain to see, is abolished. For what number of people who have already existed and will exist is it to which it is not possible to add? Also the unlimited progress of time is abolished: with this the everlastingness of the universe, which was demonstrated by both Plato and Aristotle by many exact arguments, is abolished.[250]

But again, if someone were to say that the unlimited exists, the difference between up and down is abolished, and that between the middle and the extreme, and also the natural tendency and motion of bodies. For if the places to which the motion is are not determinate, neither will the motions be determinate. Also if someone were to say that the unlimited exists, whether as substance or as attribute, we shall soon learn how many and what impossibilities result. But if someone were to say neither of these, but that all the same there was the unlimited, the absurdity seems to arise immediately that there is something which is neither substance nor attribute. In any case Alexander explains 'in neither way' not as distinguished from every kind of attribute but from that which essentially belongs to something by its nature, as if Aristotle said: 'or neither as substance nor as an essential attribute, but an attribute in some other way'.

204a1 It is particularly the task of the natural scientist to examine whether there is any perceptible unlimited magnitude.

He has established by all the foregoing that the inquiry about the unlimited is appropriate for the natural scientist, both from its substrates, since they are magnitudes changing in time, and because the ancient naturalists did not discuss it incidentally but made it a principle, and yet again because in supposing the existence and non-existence of the unlimited, and also saying that it was a substance and an attribute and neither, they encounter many problems. So in these ways he has shown that the discussion of the unlimited is appropriate for the natural scientist, and reasonably adds that an account of these kinds of unlimited will be appropriate to the natural scientist because it is not about something intelligible such as the mathematical or perhaps something superior to this, but about that which is perceptible. For it is characteristic of the wise man to make his discussions appropriate to the hypotheses before him.

204a2 So first we must distinguish how many senses 'unlimited' has. [In one way it is what cannot be traversed because it is not of a nature to be traversed, in the way that a voice is invisible; in another way it is what can be incompletely traversed, or with great difficulty, or what is of a nature to be completely traversed but is not or has a limit. Also everything is unlimited which is so] through addition or division or in both ways.

He has said with regard to what kind of unlimited it is the task of the natural scientist to examine whether it exists, that it is with regard to a perceptible magnitude, and it is this that he will also prove not to exist. But he maintains that certain things signified by 'unlimited' exist, since he holds that 'unlimited' has a variety of senses, being one of the ambiguous terms; so he first distinguishes its significations. For we can thus learn in which of its senses the unlimited is impossible, in which possible. So in common use it is what is impossible to traverse and endless that is unlimited which one cannot travel through; within this, one sort is said to be unlimited because it is wholly of a nature not to be traversed or travelled, like a point and in general everything that is not a quantity. For what is not a quantity cannot be traversed at all and therefore cannot be incompletely traversed. In that way a voice is said to be invisible negatively as not being of a nature to be seen, since it has no colour. This negation through not being of a suitable nature is made in many ways. For a wall and a fish and a swift lack feet in different ways, the first because it is not of the genus to have feet like an animal, the fish because of its difference from things in the same genus, since of animals or the aquatic[251] some have feet and some do not, the swift (this is the bird that they also call the *kenkhris*) because its feet are small and poor. But we are not inquiring whether there is a perceptible unlimited in that negative sense of 'unlimited'.

The intraversible unlimited in a second sense is set out as that which can be traversed, because it is a quantity, but one which cannot be completely traversed and is unlimited and without an exit. There are different sorts of this as well; it may be constructed in that way, sometimes through its shape, as a ring which has no bezel (*sphendonê*)[252] is traversed without limit, because every circle is. But it may be because it has no exit, as was the cavern of the Lacedaemonians into which they threw the condemned,[253] or because of enormous size, which is the most frequent concept of the unlimited.

The third sense of 'unlimited' is that of what is traversed only with difficulty, either because of size or because of construction, like the Labyrinth and the cavern later on. For the fox that led the Messenian Aristomenes out through it showed that it was traversed only with difficulty.[254]

The fourth signification of 'unlimited' is of that which is of a nature to have an exit but which no one can traverse. Such is a route through excessive heat or excessive cold, which can be traversed in that its magnitude is determinate, but is blocked through the unsuitability of the weather. Alexander put the labyrinth as well in this class, which was so constructed as to be without an exit, and the unlimited ring, as being traversable by their own nature, since they are limited, but not traversable because of the type of their construction. But perhaps the labyrinth had an exit, even if only with difficulty, and should be classed with the traversable with difficulty, and the ring, just as every circle is not of a nature to have any limit, not because of its magnitude, but because of its shape.

The fifth signification of 'unlimited' he assigns to that which is so by addition, which is observed in the case of number (for it is possible to add to every number proposed), and that by division, since every continuum is divisible without limit. But, in the case of numbers,

being unlimited is only by addition (for division halts at the unit and
goes on no further without limit), whereas in the case of magnitudes
it is both by division, when any that one chooses is divided, and by
addition also, when it is both divided without limit and if the segments be added back. This is why he said 'through addition or division
or in both ways'. In that way future time is unlimited by addition and
past time by division. But he added 'everything' not because all that
is called unlimited is so but because all that is strictly said to be so is
so, not because the simultaneously unlimited is so (for it will be shown
that there is nothing such either in number or in magnitude), but
because that which proceeds without limit is so. This is that which
has its being in becoming.

Chapter 5 begins here in modern editions.

204a8 So it is not possible for there to be a limitlessness separate
[from perceptible things, being itself something without limit.
For if the unlimited is neither a magnitude nor a number, but
itself a substance and not an attribute, it will be indivisible,
since what is divisible will be either a magnitude or a number.
But if it is such it is not unlimited, except in the way that a voice
is invisible. But those who say that there is the unlimited do not
say that it is so in that way, nor are we inquiring into that,] but
into what is so as untraversable.

The prior problems about the unlimited would be whether it exists or
not and thus whether it is a substance or an attribute; but Aristotle
right at the beginning of the inquiry inquires whether it is a substance
or an attribute, perhaps as a short cut. For if it be shown that it is
neither a substance nor an attribute of anything natural it will be
shown that the unlimited altogether does not exist as a reality. For
every reality is either substance or attribute.[255] So he shows that the
unlimited is not a substance as follows: if it is a substance it will be
indivisible, but if it is indivisible it is not limitless; but it is presupposed that it is unlimited; therefore it is not a substance. Also he
shows that substance is indivisible as follows: if it is a substance it is
not an attribute; if it is not an attribute it is not a quantity; if it is not
a quantity it is neither a magnitude nor a number; if it is neither a
magnitude nor a number it is not divisible, since these are the
divisible; but if it is indivisible it is not unlimited, for the unlimited
is divisible; but it is presupposed that it is unlimited; therefore it is
not a substance. But if, being a substance, it were said to be indivisible, it is so said as what is, by mere negation, not of the right nature,
as a voice is invisible. But the unlimited now under investigation is
not such in that sense, but as what cannot be traversed. He says that

substance is separate in the sense that it is self-subsistent and not in something else as is an attribute.

204a14 But if being unlimited is merely attributive[256] **[it could not be an element in existents** *qua* **unlimited, in the same way as being invisible could not be of speech,] although voice is invisible.**

The commentators state that 'attributive' was said as equivalent to 'attribute' so that he will be saying 'but if the unlimited is not substance but an attribute it would no longer be an element in things', as the Pythagoreans said, and Plato, and still more the natural scientists who say that the element is water, or air, or fire, or the intermediate.[257] For that which has the attribute of being without limit could be an element and a principle, as in the case of voice. For that is a principle and element of speech, and being invisible is an attribute of it; and nobody would say that this was a principle or element of speech. So, as I said, just about all the commentators interpret the passage in this way. But perhaps 'attributive' is not equivalent to 'attribute', but Aristotle was using the word strictly and had not yet moved on from substance to inquire whether being unlimited is an attribute; instead, after showing that the unlimited is not a substance as such, and that it is not the same thing to be unlimited and to be a substance, he is showing that the unlimited is not even incidentally characterized as a substance, which one might suspect, seeing that quantity belongs to natural substance. So he is saying that if the unlimited is incidentally a substance the unlimited could not be an element of things, but the substance which incidentally belongs to the unlimited, as in the case of voice and speech. For something invisible is incidentally characterized as voice since something invisible happened to be voice, but voice did not happen to be something invisible, for it is such of necessity; so the invisible is not an element of speech, but voice, which incidentally is something invisible, is. For if voice, too, is invisible it is clear that it is so among others, so that something invisible is voice. So voice is something included incidentally in the invisible in the way that substance is in the unlimited. For it is clear that being invisible is analogous to being unlimited. I know that it was also possible to interpret this as was previously stated, for as he goes on he calls attributes (*to sumbebêkos*) attributive (*kata sumbebêkos*). But the fact that he sets out still more arguments to show that the unlimited is not a substance, but has changed over to the attribute where he says 'so being unlimited occurs attributively',[258] leads me to this interpretation. Also there too he refutes this as well from the fact that it is not a principle. It is clear also from the fact that he uses in a way the same argument that the supposition here is not that of the attribute (for he would not have said the same thing twice), but that of substance belonging to the unlimited attributively or incidentally, after the supposition that it did so as such.

204a17 Also how is it possible [for there to be something which is the unlimited itself, if it is not also a number and a magnitude of which being unlimited is an internal affection? For it is still less necessary] than for a number or a magnitude to be so.   35

Alexander agrees that this also is a proof that the unlimited is not a self-subsistent substance. The argument has the form of 'the more   473,1
and less'. For if number and magnitude, of which being unlimited is an affection, cannot be self-subsistent, i.e. substances, since they are quantities and in a substrate, then still more this cannot hold for being unlimited. Therefore the unlimited is less a substance than magnitude and number. For if it were such it would be in a substrate.   5
For a substrate is less compelled to be in something else than an attribute of it, since that which is a substrate of something is more capable of being self-subsistent than its attribute; but if it is not self-subsistent it is not a substance either.

204a20 But it is also plain that the unlimited cannot be a substance [as actually existent,[259] and as substance and principle. For any portion of it you take will be unlimited if it has parts; for the unlimited and being unlimited are the same if the unlimited is a substance and not predicated of a substrate. Therefore it is either indivisible or divisible without limit. But the same thing cannot be many unlimited things; but yet the unlimited must be a part of the unlimited, as air is a part of air, if it is a substance and a principle. Therefore it cannot be shared or divided. But the actually existent unlimited is impossible;] for it must be a quantity.   10

This argument also is destructive of the unlimited being a self-subsistent substance. For he first showed that if the unlimited is a substance and not an attribute it is without parts and indivisible, and therefore not unlimited, since every substance is either divisible or indivisible, and said that it follows from its being indivisible that it   15
is not unlimited; he now draws the absurd conclusion from the supposition that it is a divisible substance. For each of its instances will be unlimited. The whole argument is as follows: if the unlimited subsists as a principle and substance in actuality the unlimited must be the same as being unlimited, since its essence is unlimited. Therefore it is simple. But if it is simple it is alike throughout. If so   20

it is either indivisible and therefore not unlimited or altogether divisible, and it will be divided into unlimited parts; for it will be into like parts as are homoeomeries. So the same thing will be many unlimited. However, even if it is not the same, still it is altogether impossible that there should be many unlimited bodies. So, if it is not a substance either as indivisible or as divided, it is not a substance.

25 In this passage he called something actually existent if it was a substance in such a way as actually to be a particular thing, capable of being self-subsistent and not as of such a sort as compounds which contain what is predicated of the substrate as one thing and the substrate as another, for these are not simple; nor did he so call them as attributes of bodies, for these also are not simples, being mixed up together with their substrates. For white is one thing, being white is 30 another. Rather he called it such as simple. For as such an existent it might be a principle, as they require. Or 'actually' was said in opposition to 'potentially', since the unlimited will be shown to be potential. So he said 'as actualized' to distinguish it from the potential, which nothing prevents it from being because of the unlimited process. But he also contrasted that which actually exists from that 35 which is merely conceived. So he says ...[260] But perhaps he contrasted 474,1 the actually existent unlimited body also to matter, which is unlimited but is not a body actually, but potentially. But he used the word 'substance' twice; once when he says 'it is not possible for the unlimited to be a substance (*ousia*)', a second time when he says 'as a substance and principle'. The first time he said it as equivalent to 5 'some nature' such as to exist actually; for the word 'existent' (*ousia*) is predicated generally of everything, as we say that a quantity or quality and generally an attribute exists.[261] But the second time he was speaking of substance strictly. But perhaps it is better to understand both as referring to substance strictly; for having said 'as actually existent' he added 'and as a substance' in order to make clear what he meant by 'actually'.

10 He said: 'if the unlimited is a substance and not predicated of a substrate' because if someone were to say that the unlimited is a number or a magnitude or in general something present in a substrate it will not follow from that that the parts of such an unlimited thing would be unlimited. For he supposes another nature for the unlimited, and nothing prevents a part of this from being a number but not unlimited; for it is not the same thing to be a number and to 15 be unlimited. For a number is a multiplicity composed of units, while the unlimited is an untraversable quantity. But if the unlimited is simple as a substance and a principle and if for all simples 'this' is the same thing as 'to be this', like the soul and being a soul,[262] whereas in the case of compounds the compound is this but being this is in accordance with the form, he who says that the unlimited is a

substance in this way, as having its being through the account of the unlimited, says that as the parts of a substance are substances so the parts of the unlimited are unlimited. For as a part of air is air, and of water water, and every part of a simple is like the whole, so a part of the unlimited is unlimited, and perhaps the parts of a part.

He said that 'it is impossible for there to be many unlimited things that are the same' not in the sense that it is possible for one thing and another to be unlimited, but because the former absurdity is more obvious. It is absurd not only because things unlimited would be many and the same but also because it seems that a limited portion is taken away from every unlimited thing and not an unlimited portion. Also Eudemus brings out another absurdity for those who suppose the unlimited to be a substance and a principle: 'for if the unlimited is an element and something is changed from it there will be something limited in what comes about; for not all things that come about will be unlimited. So it will result that the unlimited will be in the limited.'[263]

204a29 So the unlimited will be present attributively. [But if it is thus it has been said that it cannot be a principle, but that of which it is said attributively will be, air or the even number. So those who say what the Pythagoreans say would state an absurdity;] for they simultaneously make the unlimited a substance and divide it up.[264]

He has shown that the unlimited is not a substance, whether as indivisible or as divisible. Next he shows that it also is not an attribute, using the same proof which he also used when supposing that the unlimited was as a principle incidentally a substance. For he says again that if it is an attribute it is not possible to say that it is a principle and an element, but rather that of which it is an attribute, whether it be air as Diogenes said, or the even number as did the Pythagoreans.[265] So it is not the unlimited that is a principle, but that in which the unlimited is present. So whether the unlimited is an attribute of substance, or substance is incidentally unlimited, on both these different accounts substance is a principle and not the unlimited. So they speak absurdly who say that the unlimited is a principle, as do the Pythagoreans. For it cannot be a principle as being an attribute, as has now been shown; and if the Pythagoreans suppose it to be a substance they are compelled to suppose that it has parts when they say that the unlimited is the even. But it has been shown that it is impossible for the unlimited to be both a principle and a substance with parts, because it will have to be divisible into unlimited parts. In any case the unlimited is not a substance if it is the even. For the even is a number, while a number is a quantity and not a substance. But also it cannot be divisible into similar parts, as a principle should be. For the parts of an even number are not even in every case.[266]

204a34 But perhaps this inquiry is general, [whether there can be an unlimited also in mathematical objects and objects of intellect that are[267] without any magnitude; but we are making a survey of perceptible things which are our subject-matter, to consider whether there is or] is not a body unlimited in its increase.

He has recalled the Pythagoreans and wisely noted that they said that the limit and the unlimited were the common principles of both intelligible and perceptible things, and of the objects of thought and mathematics that were intermediate between them, as being the causes of union and separation. But we are doing natural science and surveying objects of perception to see whether there is or is not a body that is unlimited, not as being divisible without limit, for that is also a property of mathematical objects, but as increased without limit, which would occur only among natural objects, if it occurs at all. He said 'things that are without magnitude' also about objects of intellect,[268] referring by 'without' to being without anything natural or mathematical. It was said also earlier that it is a characteristic of the man of understanding[269] to confine his discussion to things relevant to his topic.

But Alexander says that this inquiry is called general because it is possible to use it not only in natural science but also in other inquiries. For, if there is among ideas that which is actually infinite, it will be so either as a substance or as an attribute. But one should observe that such a division is not suitable for ideas, since Plato clearly says that the ideas are substances, and does not allow that what are *here* attributes of something else, such as justice, prudence and knowledge, are *there* in something else. For he says in the *Phaedrus* (249D) that the soul carried up to the intelligible world 'sees justice, sees prudence, sees knowledge not as in becoming nor as something existing in another'.[270] In general Aristotle said 'general inquiry' not with regard to whether the unlimited is substance or attribute, but to whether it can occur also 'among mathematical and intelligible objects', as the Pythagoreans said.

204b4 If we look at it in general terms[271] it would seem not to exist on such grounds as these: [if the account of a body is that it is bounded by a surface, a body could not be unlimited, whether conceived or perceived. Nor indeed could a number be thus separate and unlimited. For a number or what has a number is what is enumerable. So if it is possible to count what is enumerable,] it would also be possible to traverse the unlimited.

He has shown that the unlimited is neither a substance (for it is neither divisible nor indivisible) nor an attribute (for it would not be a principle). He has also objected to the Pythagoreans, because they also say that the unlimited is a substance, since they say that substance is a principle which has parts, because it is the even. He next insists that such an inquiry as also examines the Pythagorean hypotheses about the unlimited, which say that the unlimited is the even, and which tests whether it is divisible and indivisible and whether it is a principle or not, is common and general and able to apply to mathematical and intelligible objects. For he says that it applies also to these cases because the unlimited is either a substance or an attribute and, if a substance, either divisible or indivisible. So he again recalls the discussion to the examination of perceptible and natural objects, whether or not the unlimited occurs in them, and first still argues in general terms, that is to say persuasively and from accepted opinion, and also more generally and dialectically. For Aristotle's dialectic is a common method for arguing about every proposed issue from accepted opinion, as he himself says at the beginning of the *Topics*.[272] For he was accustomed to contrast argument in general terms, as commonly applicable, with the appropriate which is related to the nature of the matter and with the apodictic. He showed that this proof is still in a way more general by saying 'a body could not be unlimited, whether conceived or perceived'. Also he seems to have distinguished the natural from the common inquiry and set out this argument as an example of the common, but next to conduct the inquiry which is suited to nature, when he says: 'they seem to engage rather in natural science ...'.[273] For the division into simple and compound would seem to be appropriate to natural matters. But this could be a second distinction which also itself shows that there is nothing without limit. For as he earlier showed that if it is neither substance nor attribute the unlimited does not exist at all, and as he will soon show that if it is neither simple nor compound it does not exist at all, so now he shows that if they are neither magnitude nor number the apparent unlimited things do not exist at all. So the whole order and conduct of the discussions before us seem to me to be of this sort.

Among the arguments to confirm the existence of the unlimited was one derived from our thought and imagination, because these can always add to magnitudes and numbers, whether in nature or in mathematics;[274] so he seems to put forward against that this argument in general terms. For it says that neither magnitude nor number is unlimited. For if, because we can always think of an addition to both magnitude and number, they are for that reason unlimited, it is clear that both magnitude and number should be limited, since every magnitude is thought of as being limited by surfaces together with being thought of as a magnitude, and every number which we take as existing can be counted and traversed.

Also in the case of body the argument has the same conclusion: if the definition of a body is as being that which is limited by surfaces, but the unlimited is not limited by surfaces (that he omitted as obvious), then no body is unlimited. Also the argument is common both to natural and to mathematical bodies. For it is posited *qua* body, i.e. three-dimensional, not *qua* natural. Also the argument in the case of number is common to both perceptible and mathematical number, and runs as follows: number ennumerable, for there is no separated number, as some think who abstract what is common to the concepts, but it is in the things that partake in number. Therefore it is enumerable. But the enumerable can be counted, and what is such can be traversed and what is such is limited. So number is limited and not unlimited, as imagination seemed to say. So even if it is possible to increase whatever number is taken, still the number taken is limited. For even if someone were to say that number is not enumerable, still what has a number is enumerable and limited and is such through partaking in number. For whatever number includes it makes countable, just as measure makes it measurable. But if it were unlimited it would not make that which partakes of it limited through partaking; but number is either limited or a limit. But perhaps number itself is enumerable and traversable. For the number three consists of three units, and because of this he said: 'for number or what has number is enumerable', using the alternative 'or'. Also nothing prevents unlimited progression in number by addition and in magnitude by division and addition, but the self-subsistent and actually unlimited is in neither of these.

204b10 If we look at it more from the point of view of natural science its non-existence is clear from the following: [it cannot be either a compound or simple, so the unlimited will not be a body if the elements are limited in number. For they have to be many, and the opposite ones must be equal in number and one of them cannot be unlimited. For if the potentiality in one body is inferior by any amount to that in another, e.g. if fire is limited, air unlimited but an equal quantity of fire is so many times greater in potentiality, but only by a certain amount, it is nonetheless clear that the unlimited will be superior and destroy the limited. But both cannot be unlimited; for body is what is extended in all directions and the unlimited what is extended without limit,] so that an unlimited body will extend everywhere without limit.

This is not in general terms and in common, but rather demonstrative and appropriate to the things before us. For his proof is from their essence, if the unlimited can be neither compound nor simple, and from the potentialities which natural objects naturally changing have, and from the places in which they naturally remain or to which they naturally move.

The proof from their essence is that an unlimited natural body can be neither compound nor simple; for it is a property of a natural body to be simple and compound. So he says that if some natural body is completely unlimited it must be either compound or simple; and, if it is compound, it must be compounded of many elements (which is why he himself added 'for they have to be many') which, too, are different in kind. But these quantities must be either limited or unlimited in number. (He did not omit this division, as Alexander thinks, because he has proved and will prove it, but will also himself examine it where he says 'but if they are unlimited and simple'.[275] For he showed that the element is not single and yet, for the completeness of the proof, he also supposes the unlimited to be simple.) So, if it is a compound of elements limited in number, either all are limited in magnitude or all are unlimited, or some one, some the other. But if they are all limited in magnitude, then, since they are also limited in number, the whole will also be limited in magnitude and not unlimited. But he excused himself from examining this section of the division as being clear. But if some are limited, others unlimited, or even a single one of them is unlimited (for the proof only needs one case), then the unlimited one transforms them all into itself. Also the elements will no longer be many, but one, nor will the unlimited be composite but simple, which he will prove a little later.[276] For the elements must be equal in number and in strength, if they are to be preserved. But if air, for example, were unlimited but fire limited, the fire would be overcome by the air and transformed into it. But perhaps someone will say that, if the greater quantity of air is weaker than the lesser quantity of fire, what is to prevent the limited fire not being defeated by the unlimited air? So he resolves this objection by saying that if an equal quantity of fire is stronger than an equal quantity of air, and if it has the superiority of power by some number and ratio or another, such as ten thousandfold, then if we take ten thousand times as much air initially it will be of equal strength with the initial quantity of fire. And if we thus take away from it many times, the fire, which is *ex hypothesei* limited, will be used up, but the air, which is unlimited, will not be used up, so that by exceeding in power it will destroy the fire. But it is clear that if the unlimited should be more numerous, the rest will be still more overcome by the more numerous.

But if all the elements should be supposed unlimited in magnitude while being limited in number, then, since body simply is what is three dimensional, extended in all directions and an unlimited body is what is extended in all directions but without limit, it is clear that if there be one unlimited body there would not be another body in addition to it. For it would be limited by it. But the impossibility follows still more if all bodies limited in number but unlimited in magnitude were supposed to be unlimited. Perhaps for this reason he solved the former problem in terms of powers and this one of extensions, though able to reduce this absurdity also to the previous supposition which took one or more but not all to be unlimited in magnitude, since this absurdity follows still more from that hypothesis.

But if someone were to say that the simples from which the supposed unlimited compound was composed were unlimited in number, which is the remaining cut of the division, that supposition was already refuted in the First Book, where it was shown that the elements are neither one nor unlimited,[277] but more than one, limited in number and not unlimited. He will soon disprove this hypothesis also, when he says 'but if they are unlimited and simple',[278] from the fact that the potentialities in accordance with which both natural and simple bodies are given their form are determined by their forms. For the primary elements are informed by the primary oppositions of simples, which are two, heat-cold and dry-moist. But things whose forms are limited in number must themselves have a limited number of differences.

204b22 But, indeed, nor is it possible for an unlimited body to be one and simple, [either, as some say, that which is in addition to the elements, from which they are generated, or simply. For there are those who make the unlimited thus rather than air or water, in order that the others should not be destroyed by the one of them that was unlimited. For they stand in opposition to each other, as e.g. air is moist,[279] water is cold, fire is hot. If one of them were unlimited the others would have already been destroyed. But now they say that there is something else from which these come. But it is impossible for there to be such a thing. Not because it is unlimited; on that something general must be said applying equally to all of them, air, water and anything; it is because there is not such a perceptible body apart from what are called elements. For everything is resolved back into that from which it came, so that it would be here in addition to air, fire, earth and water; but nothing can be seen. Nor can fire or any other of the elements be something unlimited. For altogether, apart from one of them being unlimited, it is impossible for the universe, even if limited, to be or become one of them, as Heraclitus says that they all at some time become fire. The same argument applies to the one element which the natural scientists posit in addition to the elements; for everything is transformed from opposite to opposite], e.g. from hot to cold.

He has shown that no natural body composed of many elements can be unlimited. He next shows that also not even a single and simple body can be unlimited. For if it is simple it will be one of the four elements or something else in addition to them, as the associates of Anaximander affirm that one in addition to the elements from which the elements are generated. It is clear that none of the elements can be unlimited also from what Anaximander said; he held that the element was unlimited, but did not make it air or fire or one of the four elements because of their relation of opposition to each other, and because if one of them were unlimited the opposites would have been destroyed by it. He also shows that the alleged unlimited as a principle beside the elements is not only not unlimited but does not exist at all, taking as premiss that everything that comes to be from something evidently is resolved back into it. So, if coming to be were from this, there would be something here beyond the four into which the resolution was occurring. But nothing such can be seen here. But perhaps someone will say that also according to him the compound is of matter and form, so that their decomposition is into these. And yet matter never appears here naked, nor form on its own. But perhaps the cause of this is that matter always possesses a form that is different at different times, and coming to be is from opposed forms and ceasing to be into opposite forms. So, since the enmattered cold came to be from the enmattered hot, when the enmattered cold ceases to be it is converted into the enmattered hot, and this is visible. So even if it be said that what comes about does so from matter and the opposite form, still it is from matter that is always present in it and from the opposite form as being transformed into the opposite. It seems that Eudemus also, having raised this problem, solves it in much the same way. For having said that here nothing can be perceived beyond the Empedoclean elements and their compounds he says: 'but it is possible for those who maintain the formless. For that is a substrate for the others.'[280]

So he has shown that the unlimited is not any one of the four elements, nor is it anything at all beyond the elements. Next he shows by a general proof that no body among the simples or the elements, nor anything beyond the elements, is unlimited. He shows first superfluously that it is wholly impossible for there to be one element of things. But if altogether there is no single element it could be neither unlimited nor limited. For Heraclitus said that everything comes from a limited quantity of fire and that everything is decomposed into it.[281] The Stoics also should be of this opinion; for the conflagration hints at something such, and they say that every body is limited.[282] He says that it is impossible, even if somebody says that the universe is limited, that there should be one element of it and some one nature, so that what exists is this alone, or that things should come to be from this nature and be transformed into it, as Heraclitus holds, who says that everything turns to fire and comes from fire.[283]

He shows that neither one of the four elements nor that which is beyond the four is the only element of things by recalling the proof in Book I,[284] by which he proved that everything transformed is transformed from opposite to opposite, so that that which is beyond the four could not be an element, because it is opposed to nothing. So it would neither be transformed into anything nor would anything come from it, whether an element or anything else. If so, it is not an element. But it cannot be any one element of the four, for if anything were to come to be from it, it will have to be transformed into it and clearly also into its opposite. But if it were transformed it would also perish into that into which it is transformed, and would come to be from it. For it comes to be from that into which it perishes. If this is so, no one of these could be the element. For why would this be the element of the others that came to be from it rather than their being so from which this comes to be? So there must be opposite principles that co-exist. But if someone were to say that, while preserving its opposition in which its being lies, it is transformed into its opposites, then it would simultaneously be in two opposite states, which is impossible; also opposites will hold of a single element. Further, if transformation is into an opposite and there is one element of all things all things will be opposite to the one, in order for it to be able to be transformed into them. But one thing remains opposite to one thing. And also there will not be coming to be, if coming to be is through transformation into the opposite. But the one is not among opposites, since the element is one and simple.

One should observe that those who say that there is something else besides these four which is the element of things treated it as matter and did not see in it opposite qualities. Therefore they also say that the four elements come from it and that it is given form by their qualities. And those who say that there is the one principle of these elements do not call it a principle as being transformed but as being seen in all things, like matter. Therefore they posit things that easily take an impression as the principles of the elements. But it is clear that it would produce the removal of composites from that being itself which stretched over all things, if it were unlimited, as also of transforming powers, and produced the removal of the simple from natural powers among which there is opposition.

It is worth pointing out about this text that among examples of elements he said that air was cold and water moist, which is not what he usually said. For he calls air moist as having more of the character of moistness which is difficult to define with its own definition but easily distinguished from something different. So perhaps here also he did not trouble because he was using the example only on account of the opposition, or perhaps there is some scribal error.[285]

205a7 But one must observe generally from these considerations [whether it is possible for there to be something unlimited.[286] It is clear from them that it is altogether impossible for there to be an unlimited perceptible body. For everything perceptible is of a nature to be somewhere, and each has a place, and the place of part and whole is the same, as of all earth and a single clod, of a fire and of a spark. So if all is of the same kind it will be unchanging or continually moving. This however is impossible. For why rather downwards or upwards or anywhere else? I mean, for example, if it be a clod where will it move or where will it rest? For the place of the body of the same kind as itself is unlimited. So will it occupy the whole place, and how? So what or where will its motion or rest be? Will it rest everywhere? Then it will not move. Or will it move everywhere? Then it will not stay put. But if the universe is diverse, so will the places be. Also, first, the body of the universe will not be one, save by contact. Next, the bodies will be either limited or unlimited in kind. But they cannot be limited; for some will be unlimited and some not, if the universe is unlimited, e.g. fire or water; but the unlimited involves destruction for its opposites, as was said before and below. And for that reason nobody among the naturalists made fire or earth the one unlimited element, but either water or air or the intermediate between them, because the place of each was clearly distinguished, and these vacillate between above and below. But if the elements are unlimited and simple then so are places unlimited and the universe. For it is impossible not to make place and body equal. For place is not greater than the amount of body there can be. At the same time, body will no longer be unlimited. Nor is body greater than place. For either there will be some void] or body of such a kind as to be nowhere.[287]

He has said and demonstrated at one time that there cannot be any compound natural body that is unlimited, and at another time that it cannot be a simple one. Now he proposes to produce a general proof about every sublunar natural body, both compound and simple, whether it is or is not possible for it to be unlimited, dividing body up into the uniform and the diverse, as earlier into simple and compound. As previously he constructed his arguments on the basis of substances and their natural powers, so now he will look from the perspective of natural places. He now constructs the discussion about a body that comes to be and ceases to be, showing that such a body is not unlimited. For he will prove of the divine body with circular orbit that it is not unlimited in *De Caelo*.[288] He presupposes[289] certain axioms about every perceptible body that was generated and is destructible. For the present discussion is about them, as the argument from the difference of the motions suitable to sublunary elements makes clear. So every such body is in place, however place be taken, whether as an interval or, as Aristotle thinks, the limit of the container. For if place is an interval, then also every natural body is in a place. But if it is the limit of the container, not every body is in a place, for the fixed sphere is not unlimited as well.[290] So that is why Aristotle conducts his discussion about the sublunary when using this axiom that every perceptible body is in place. The second axiom is that each of the natural bodies has some place of its own, and the third is that in the case of things with similar parts (homoeomeries) the place of the part and the whole is the same; for example that of all earth and one clod and fire and a spark. The place is the same in kind, but not so in quantity. For if the whole earth were to move, it would move to the place to which the clod moves; and where the earth remains, there its parts also naturally remain.

So he makes these presuppositions, and also that every natural body has either similar or dissimilar parts, this being a further division beyond that which divides them into the simple and composite. For a compound can have similar parts, like flesh and bone and similar things. And in any case it would not have been worthy of Aristotle's breadth of view to use the same division once again. Also what is said is not said as about the simple and composite, but about things of like and unlike parts, which he calls 'of the same kind' and 'of different kinds'. Even if he calls it 'simple' where he says 'if they are unlimited and simple', he sets it down on the ground that the simple is more akin to that with similar parts, because of the comparison with the organic that are compound. So we beg to differ from Alexander, who here treats 'simple' and 'of the same kind', 'compound' and 'of different kinds', as identical and says that Aristotle has now made the same division as he made before, when he made the division into simple and compound.

So with these posits he shows that there is no natural body, whether with similar parts, i.e. of the same kind, or different parts, i.e. of different kinds, which is unlimited, so that there is absolutely none, since every natural body has either similar or different parts and neither of these can be unlimited. First he shows that no natural body with similar parts is unlimited. For, if it is, let some part of it be taken, e.g. if it be earth, a clod. So, since the place of a part and the whole of things with similar parts is the same, where will the clod travel or where will it remain? For the place of an unlimited thing is unlimited, as will be shown. So if the earth is unlimited and its place is unlimited, and the same place is proper to the clod as well, where will it go, up or down? For the whole is equally proper to it, so not rather here than there.[291] But if everywhere, since every part of place properly belongs to it it will be torn apart as it travels everywhere at once. But, if it will not move, where will it remain at rest, since it has no assigned place? So it will not remain. So it neither remains nor moves. For if it remains at rest in the whole it will not move, and if it moves in the whole it will not rest since the whole place is unlimited and every place is proper to the clod because the places of the part and the whole are the same in kind, even if different in magnitude. But if it will neither move nor remain at rest it will not be natural since it has not within it the nature which is the principle of motion and rest. So thus it is impossible, and it is in any case absurd that there should be a body which can neither rest nor move. So the existence in this way of a simple unlimited body with similar parts is eliminated.

But also it is not possible for it to have dissimilar parts, i.e. of different kinds, if it be supposed to be unlimited. Here also he presupposes it as obvious that, if the whole is of different kinds, the places will also be dissimilar. For its proper place is assigned to each sort of body. But if both the bodies and the places are different the unit will not be continuous but made of parts touching each other. But he did not adduce this as absurd, for as things are the single constitution of the universe is out of touching parts and not of continuous ones, as the different kinds of motion show. Rather he made this point in order to establish that the parts of that with different parts are many in actuality. With this presupposed, he next makes use of this division: if the parts are many and differing in kind they will either be limited in number or unlimited; if they are limited in number but the whole is unlimited in magnitude it is again necessary that one or more should be unlimited. For if all are limited then the whole will be limited and not unlimited. But if one of them is unlimited the previously stated argument will arise, that the others will be destroyed by the unlimited one which is different both in form and underlying place. But the differences of place result also in mutual oppositions, both of the places themselves and of the bodies in the places to one another. Also he shows that there cannot be one body that is unlimited if places are distinguished, both since there would at once be opposition and destruction of the rest, and he also shows this through the supposition of others, the naturalists who said that the element was single and unlimited. For none of them said that the element was fire or earth, because these, if they had distinguished places and were unlimited, would have destroyed utterly those among the opposites which had an opposite nature. But they said it was water, as Thales did, or air as Anaximenes did, or that intermediate element, as Anaximander did, because these seem to be ambivalent in their places and can be above or below and thus were not destructive since not opposed in their places. Also it is in any case easier to interpret them so that none located the one unlimited element in a distinguished place. For a distinguished place would not be the whole, but the place of the unlimited would be the whole of it. One would not be showing the impossibility of the supposition from the destructive character of the unlimited but from the premised distinction of places. It is clear that the supposition that makes one of many unlimited is impossible for another reason; for how will it give ground for the existence of another or how will it not be limited by it? But Aristotle himself omitted to make this refutation, as being clear.

He went on to the other section of the division which supposes the parts to be unlimited in number. (He says that, if they are unlimited and simple, that means that they have similar parts, for that which is the first with different parts is composed throughout with nothing but parts with similar parts, as for example our hand, which is composed of flesh and bones and sinews.) So if those parts which compose the unlimited whole with different parts are unlimited in number and in kind, their places will also be unlimited and the differences of the places. But if the places and their differences are limited, as he will demonstrate, then the simple bodies will be limited as well, in kind and in number, so that the whole will also be limited in magnitude. Also it is clear from the oppositions of place that the presupposition is true and places are limited; for there are six differences of place: – above and below, on the right and on the left, in front and behind, and none in addition to them. He demonstrates also the conditional proposition that if the places are bounded then also the bodies in them will be bounded and limited from the need for the total equality of place and body. For if there is an excess of place two impossibilities will follow; both that there will necessarily be a void, which will be proved not to exist [in Book 4], (for those who say that there is a void say that it is a place deprived of body)[292] and also that, if place were greater than body, body would not be unlimited. For nothing is greater than the unlimited. But if someone were to say that there was an excess of body over place there will be a body which is not in a place, which does not please the majority of scientists who take all body to be in a place.

'And the elements will be unlimited' could be said with reference to the supposition from which is inferred 'but if that is impossible'; he would thus be calling elements the simples with similar parts which were parts of the unlimited whole with different parts. But it may itself be said as an absurdity consequent on the supposition. For since the unlimited whole with different parts consists of simples unlimited in kind according to the supposition, these will be its elements. But it was shown in the First Book that the elements cannot be unlimited in number.[293] But as it is it looks as if 'but if this is impossible' corresponds to the first interpretation, as the cause of the impossibility inferred from places makes clear.

205b1 Anaxagoras speaks absurdly [about the unlimited being at rest. For he says that the unlimited supports itself. This is because it is self-contained (for nothing contains it), since a thing naturally stays where it is. But this is not true; for something might be somewhere by force and not where it is naturally. So if the universe does not move at all (for what is self-supporting and self-contained must be unmoving), still it should be said why it is not of a nature to move. For it is not sufficient to say this and leave it there. For it might be also that it does not move because it has nowhere else to move to, but nothing prevents it being of a nature to move. For the earth would not move, even if it were unlimited, being instead stopped by being at the centre. But it would not rest there because there is nowhere else where it might move, but because that is its nature. However it would be possible to say that it supports itself. So if that is not an explanation in the case of the earth, if it were unlimited, but it is because it has weight and what is heavy remains in the middle and the earth is in the middle, similarly the unlimited would remain in itself for some other reason and not because it is unlimited and supports itself.] At the same time it is clear that any part would have to be at rest. [For the places of a whole and a part are of the same kind, as that of the whole of earth and a clod are below and those of all fire and a spark are above. So if the place of the unlimited is within itself the place of a part of it is the same. So it will remain within itself.][294]

He has just concluded that it is impossible "for a body to be naturally nowhere" because every body is of nature to be somewhere; also previously he took it as an axiom that there is a natural place for each sort of body,[295] and that in the case of things with like parts each part travels towards that to which the whole does, and remains in that in which the whole does. Having drawn the above conclusions on the basis of these axioms as generally accepted, he immediately finds Anaxagoras contradicting them. For he says that the unlimited mixture of things with like parts is unmoving, and says that the reason for its immobility is not that it is in its proper place; for it is not even in place at all because as unlimited it is not contained by anything, since what contains must be greater than the contained, while the unlimited must have nothing greater than itself. For this reason, then, he does not make place and its affinity with place responsible for its rest, but the fact that it is self-supporting. So Aristotle contests this opinion on the ground that Anaxagoras does not rightly think that wherever a body happens to be it is naturally there, and that for this reason the mixture is self-contained but not in place, and thus remains as it is as being naturally self-contained. For this is not true; for it might be somewhere also forcibly, like a stone hanging from a peg which stays where it is not naturally. And so if the universe does not move, being self-supporting, because it is necessary that what is self-supporting and self-contained should be immobile, this necessity is not independent, and he says that it is at times forced as an account of the natural cause of its rest. Rather we should look for a natural cause of its rest. For something may at times be at rest not because that is its nature but because it has nowhere else to go, being unlimited, though of a nature sometimes to move rather than be at rest. Thus it does not suffice to say that it rests somewhere either in itself or in something else in order to show also that it does so naturally. For the fact is not the same as its explanation. That it rests is a statement of fact, but he who is looking for the explanation of its rest is not satisfied with the fact of rest. He shows that this is the case also with the example of the earth. For even if someone supposes it to be unlimited (so that he should not believe that there is some result from this, but e.g. that it would rest and support itself), still this is not why it would rest, that as unlimited it would have nowhere to travel to; nor is it because it was self-supporting, but because it was constrained from moving by the power of the centre in which it is its nature to be. So if also in the case of the earth, even if one supposes it unlimited, the explanation of its rest is not the support, nor having nowhere else to travel to, but some other natural explanation. (It is because it has weight and the heavy remains in the centre, and so the earth remains in the middle), if this is so, then in the same way, if the unlimited mixture were to remain in itself, it would remain because of some other natural cause. And it is not because it has nowhere to travel to as being unlimited nor because it is self-supporting. He who says that it is does not state the natural explanation of its rest, but an explanation from force and necessity, nor the cause of the earth as earth but of its magnitude. It is because of this, its magnitude, which is supposed to lack a limit, that it has nowhere else to travel to.[296]

So he refuted Anaxagoras' explanation of rest in this way. But he refutes it still more with reference to parts. For if the part is of the same kind as the whole unlimited mixture, since the whole of it is in the whole, and if the places of things of the like kind are themselves of the same kind, and the whole unlimited mixture remains at rest self-contained and self-supporting, then it is clear that if any part whatsoever of it be taken it remains at rest also and is self-supporting. For it is self-contained according to its place. Where the whole is, there is the part, if it is of the same kind. For the place below is that of both the whole earth and a single clod, and the place above of all fire and a spark. So if the place of the unlimited is self-contained the place of the part is the same. Therefore the part also is at rest and will not move. This is obvious from observation;[297] for all the parts move. But if they should be in their own place they must remain at rest. Anaxagoras took 'self-contained' as meaning 'in their own place', and therefore said that the whole mixture remains at rest because it supports itself in itself. But generally if it remains at rest because it is self-contained, and if this being self-contained belongs to both the wholes and the parts, whether limited or unlimited, everything would rest and nothing would move. But being at rest does not belong to the whole mixture as being unlimited, since it also rests because it is self-contained. But it is clear that if nothing moves naturally then also nothing will move unnaturally; for unnatural motions are those which are opposite to natural ones. Therefore no natural body will move in any way, and this even though nature is the principle of change.

He has conducted the argument as applying to one part, but lest anyone should think the conclusion was drawn as applying to one part he concluded that it is at the same time clear that any part whatsoever would have to be at rest. For if it is natural for any one part of the unlimited to be self-supporting, and so to be at rest, it is the same for all. Therefore no kind of body will move naturally. But nor will it move unnaturally, as he will prove in the fourth book;[298] for the unnatural is subordinate to the natural. But this is utterly absurd and is at war with Anaxagoras' doctrine; for he does not say that by observation things are unmoving. But the inquiry whether there is something self-sufficient and whether something self-sufficient can be a place belongs to the inquiry into place and is introduced in the discussion of place at the beginning of Book IV.[299] But it is clear that with regard to the whole he refutes the argument as failing to give the natural cause of rest, and with regard to parts since the absurd consequence that nothing natural moves follows on the stated cause.

So Aristotle met the superficial version of the doctrine of Anaxagoras in that way. But if I previously[300] said truly that Anaxagoras exhibits a double universe, one intelligible, one perceptible, the former unified, the latter discrete, its not being in a place but in itself and being self-supporting will be very suitably said of the intelligible one, as also its being unlimited. For bodies limited by surfaces, separated as they were poured forth, also needed some receptacle and support, which place provides for them, since they have fallen away from themselves and from being self-contained, and have come to be in place as also in matter. For the former have no parts and are indivisible in themselves and are mutually supporting as being unified, but the latter are separate, with parts thrown apart in different places, since the whole is in different places and not in itself, and need a place and a way of ordering such fragmentation. But we must move on to what follows next.

205b24 In general it is clear that it is impossible to say both that body is unbounded and that there is some place [proper to bodies, if every perceptible body has some heaviness or lightness and if the heavy naturally travels towards the centre, the light upwards. For it is necessary that the unlimited should do so also, but it is impossible for the whole to be affected in either way or for either half to be affected. For how will you separate them? Or how will there be some of the unlimited above and some below,] or at the edge and in the middle?

Having examined the account of Anaxagoras, he rightly maintains that 'being in itself' was used instead of 'in place' as being equivalent to 'being in place', because it is consistent for one who says that body is unlimited not to say that there is some place proper to bodies and for one who accepts place to deny that there is an unlimited body. It is clear that this demonstration itself contributes to the refutation of the unlimited projected initially. For just as he showed when arguing formerly from the unlimited itself that it is neither substance nor attribute, nor simple nor complex, neither with like nor with unlike parts, so now he argues from places and its relation to them. First, that if it is unlimited it is not in place. The argument starts from the powers of natural bodies. For if (a) every perceptible body has either heaviness or lightness, and if (b) for bodies the motion from places and to places, and rest in them, comes from these tendencies, and if (c) it is impossible for an unlimited body to have either heaviness or lightness, as he will show, then it is impossible for any unlimited body to be in a place. But the premiss is true, so also the consequence. He omitted that every perceptible body has either heaviness or lightness as obvious. For universally it must have one of these contrary qualities; simple bodies must have one of the qualities without qualification, mixed ones in accordance with the mixture itself or the predominant quality. He showed that if there is some unlimited body it is impossible for it to have heaviness or lightness or be above or below by saying 'it is impossible for the whole to be affected in either way or for each half to be affected'. He did not add to these two the absurd consequence of the whole being affected in either way; for if the whole is either above or below, then, since each of these is limited if above is towards the outside and below is towards the middle, the unlimited body will be in a limited place, which is impossible. Also, if the unlimited is in either of them, the joint place will be greater than the unlimited, given that it is equal to the one in which it is. But nothing is greater than the unlimited. So he did not think it worthwhile to add these points to the argument as being utterly clear. But he adduced the consequence of each half being affected by saying: 'how will you divide them? Or how will one half of the unlimited be above, the other below?' For it is impossible to divide the unlimited into two. For either each half will be unlimited and the whole greater than each of the unlimiteds, or each half will be limited and the whole will be limited as consisting of two limited parts.

But how will one part of the unlimited be above, the other below? For the two parts of the unlimited will be in limited places. So they will themselves be limited and the whole will be limited. He indicated that above and below are limited by 'at the outside or in the centre'. For what has an outside and a centre is limited. For 'outside' and 'centre' are names of limits. So what is above will be the outside of the unlimited and what is below the middle. But if the outside is a limit, but below is in the centre, and each is limited, and therefore it is impossible for the parts of the unlimited to be in them, then it is very much clearer that it is impossible for the unlimited whole to be in either of them. So by the same proof he showed that it is impossible both for the whole to be affected in either way, and for each half of it to be affected.

205b31 Further, every perceptible body is in place, [and the different kinds of place are above and below, in front and behind and right and left, and these are not merely relative to us and conventional but are distinguished in the universe itself.] But it is impossible for these to exist in an infinite body.

He has shown that there could not be an unlimited in place from the natural powers of a natural body and its natural tendencies derived from them, i.e. from their heaviness and lightness. So, having shown that the unlimited would not be in a place, he now refutes, from a consideration of the differences of place the possibility of an unlimited body being among them. The object was to show that it is impossible to say both that body is unlimited and that there is a place for bodies. So having shown that if it is unlimited it is not in place, he now shows that if it is in place it is not unlimited. For if every perceptible body is in place it is impossible for body to be unlimited. The kinds and differences of place are above and below, in front and behind and on the right and on the left, and these are not merely relative to us and conventional, as will be proved,[301] but are naturally distinguished in the universe. So it is clear that they are distinct and limited, as their differences from each other make clear. So the natural body in them will be limited. For a body is equal to the place it occupies, as was previously shown. Therefore it is impossible for the different types of place to accept unlimited bodies, so that what is in place should be unlimited. He shows that these differences of place are natural in *Cael.*,[302] and so does Alexander in his work written against Zenobius the Epicurean.[303] Aristotle proves admirably that these could not be subjective and relational unless they were already in the universe and natural.

205b35 Without qualification, if it is impossible for place to be unlimited, [but every body is in a place, it is impossible for there to be an unlimited body. But what is somewhere is in place, and what is in place is somewhere. So if it is also not possible for the unlimited to be a quantity, since it will be some quantity, such as two or three yards, for that is what 'quantity' means, similarly what is in place is somewhere, and that is either above or below or in any one of the six directions,] and there is a limit to each of these.

He has shown from the differences of places that the unlimited cannot be in place, nor what is in place be unlimited. But lest anyone should say that the unlimited is not in a particular place, but asks what prevents there being an unlimited body in the whole and in the place that is unlimited without qualification, he answers that objection by saying that if it is impossible for a place to be unlimited, but every body is in a place, it is impossible for there to be an unlimited body. The protasis is true and therefore the conclusion. He omitted to confirm the conditional as totally obvious, but he puts in the additional assumption saying that it is indeed impossible for place to be unlimited and confirms it by showing that it is impossible for what is not in some particular place to be in place, and that this is so because the validity of general truths rests in the particular.[304] For that existent of which a genus is predicated is itself an individual; e.g. if 'a man', then also 'a particular man'. Thus what is somewhere, e.g. what is above or what is below, that is also in place and what is in place is always somewhere, i.e. above or below or simply in some particular place. For because of this the unlimited is not simply a quantity, because it is necessary for it also to be some particular quantity, such as two or three yards, if the general has its being in the particular. However Eudemus says that also 'some' has the same relation to 'this', as 'of some sort' to 'of this sort', 'a quantity' to 'this much', 'somewhere' to 'this location', and that the former are general the latter particular; e.g. 'somewhere' is 'in place', 'this location' is 'at Athens'.[305] But Aristotle seems to take 'somewhere' as equivalent to the particular.

Also I think it better to take 'if it cannot be a quantity' and 'it will be some quantity' to refer to quantity[306] in general; in 'it will be some quantity' 'quantity' should be understood as referring to some definite quantity. For the added 'some' shows that it is definite. But it is possible to read it as Alexander does; so if the unlimited cannot be a limited quantity, because thus it becomes a particular quantity such as two or three yards, then nor can it be in place, because it will be in some one place such as above or below or some other direction in place, of which there are six. But it cannot be in any one of them, because each is a limit. Thus he said also 'but indeed what is somewhere is in a place and what is in a place is somewhere', using 'somewhere' as equivalent to 'in some place', such as above or below, as he soon made clear by saying that what is in place is somewhere; i.e. either above or below. So it is perhaps better to make one 'quantity' definite (*barutonein*), but not to do so with the first but the second 'for it is some one quantity', so that 'quantity' so written will point to the definite which is some one,[307] two yards or three yards. From that it is clear that Aristotle does not abolish the existence of general terms, but says that the general exists in the many different instances and denies its self-subsistence. He thinks that it is the work of our thought to separate the generality which exists in the many. Therefore he also calls it posterior, but the transcendent, from which the immanent generality comes, would not be justly called generality.[308]

206a7 From these considerations it is clear that there is actually no unlimited body. [But it is clear that many impossibilities result if there is nothing unlimited without qualification. For there will be some beginning and end of time, magnitudes will not be divisible into magnitudes and number will not be unlimited. So, when after the distinctions made both alternatives are apparently impossible, arbitration is needed, and it is clear that] in a way there is the unlimited and in a way there is not.

In modern editions chapter 6 begins at 'but it is clear that many impossibilities' above.

5 He has previously stated arguments from which one would believe in the unlimited, and then he has shown by many arguments that there is no unlimited body in actuality. Next he makes the arguments clash, and says that if some absurdities follow for those who posit the unlimited and also for those who without qualification abolish it, there is need for an arbitrator who will show that in a way there is
10 the unlimited and in a way there is not. For if absurdities follow for one who proves that it does not exist it is clear that it must, and, again, if impossibilities result for one who proves that it exists, then it cannot exist. So if the opposing views are not both true it must be that in a way there is, in a way there is not. But since he emphasized the absurdities resulting for those who completely abolished the unlim-
15 ited, for that reason he also now recalls them and puts before us the most important. For if time is not unlimited there will be some beginning and end of time, and, if so, there was a time when there was not time and there will be when there will not be. But past and future are parts of time. Also unceasing change, of which time is the number,[300] will be abolished, and the universe will not be everlasting.
20 Second, if there is nothing unlimited at all, the division of continuous quantities will come to a halt and also the increase of number.

206a14 Now one thing is said to exist potentially, another in actuality, [and the unlimited is sometimes by addition and sometimes by division. It has been said that magnitude is not actually unlimited, but it is so by division; for it is not difficult to get rid of indivisible lines;] it remains that the unlimited is potential.

As he is about to show the way in which the unlimited exists, and the
25 way in which it does not, he makes some assumptions that will be useful to him for the proof. One is that of things said to exist some are said to do so potentially, some to do so actually; this he assumed as obvious, customary and often proved by him. Another is that there is the unlimited by addition and by division. This, too, is obvious, since it is absurd for there to be a beginning and end of time, that magni-
30 tudes should not be divisible into magnitudes, and that every given number should not be capable of increase. If, however, the unlimited must exist, both by addition, because of time and number, and by division because of the cutting of magnitudes, but no magnitude is actually unlimited, as had just been proved,[310] then the unlimited
35 magnitude must exist potentially, since everything there is is either

potential or actual. So the unlimited is potential in the division of   492,1
magnitude. So magnitude is divisible without limit.

For he also says that there is no difficulty in getting rid of indivisible lines, which seem to prevent unlimited division. For he both wrote a book against indivisible lines and will show in this work that magnitudes are not composed of bits without parts.[311] But this is also   5
clear from mathematics. For if, as was demonstrated in the sixth book of Euclid's *Elements*, it is possible to divide every line in the same proportion to one already divided, there could not be an indivisible line.[312] Also, if there are indivisible ones, not every line will be divisible into halves as he who professed to bisect any given line would seem to do. Also it is impossible not only to cut the indivisible line in   10
half but also so to cut one composed of an odd number of indivisible lines, since an odd number also cannot be divided into two.

206a18 But one should not take the potential [in the way that as, if this is potentially a statue, then it will be a statue so what is unlimited will be so actually. But since existence is of many kinds, the unlimited exists just as a day and a contest exist by one thing happening after another. For in these cases also there is both potentiality and actuality; the Olympic games exist] both through the potentiality of the contest occurring and through its actually occurring.

He has said that the unlimited is potential in the division of magnitudes in virtue of its being without limit. But since all the potential   15
also must sometimes reach actuality unless it is pointless, and if there should come to be in actuality an unlimited number of sections having magnitude, then magnitude would come to be unlimited. For a magnitude consisting of magnitudes unlimited in number will be an unlimited magnitude, as has often been stated. But it has been demonstrated by many arguments that there is no unlimited magnitude. In resolving this objection, and at the same time showing more   20
plainly the nature of the unlimited, he says that the potential is divided in two in line with the division of the actual. For since the potential is spoken of in relation to the actual, the potential will be spoken of in as many ways as is the actual. There are two ways of being in actuality; for it is either what exists as a whole as what it is, like a man or a house, or as what has its being in becoming, like a contest and a day. For we say that these also exist in actuality when   25
they exist. So therefore there are two ways of being potential, the one in relation to what exists whole all at once, as we say that the bronze is potentially a statue, because it sometimes emerges into being a statue, which exists whole complete, the other in relation to what has its being in becoming. So the unlimited in the division of magnitudes

is potential, not because magnitudes are ever divided without limit, but because their division is not limited, i.e. they can always be divided. For what is without limit is not so actually, but its being without limit is potential. For it is unlimited by being always capable of being cut and the cutting never being exhausted. For what has been cut off is always limited, but because every section can be cut again it is thereby without limit. If in the case of division and that of addition there is always something further, in the case of division that which will be divided and in the case of addition that which will be added, its being unlimited is potential. In any case even if it is not divided it is divisible. But if being without limit is seen in its being divided, do not wonder at it. For that actuality which has its being in becoming always is together with the potential, and it always is extended in coming to be for the reason that nowhere is it free from the potential. For when free from it it is what it is and has a limit and nothing without limit. For that reason things that exist only in actuality are simultaneous wholes, and there is neither any potentiality in them nor limitlessness. Therefore it is not time which changes numerically that is such, but eternity which remains unchanged in unity.

Therefore the unlimited is reasonably said to be potential and not actual. For that which has parts is always bounded, as Eudemus says, and is not without limit.[313] For a limited part is added to a limited part, and division does not come to an end. For what might come to be is not definite, but the undefined quantity is without limit. It appears likely that Plato also says that the great and the small is unlimited because the potential absence of limit is in either direction.[314] So in actuality neither number nor magnitude is unlimited. Also it seems that in the case of the unlimited potentiality and actuality are the same. For the actuality of the unlimited as such is continual potential to increase, since if one were to seek for the actuality of the unlimited, such as some stability and form, one would seek nothing other than a limit of the unlimited, that is to say its ending, which is impossible. For every actuality ought to preserve its substrate. And just as the actuality of the changeable preserving the potential is change, so is the actuality of the unlimited. Just as things having their being in becoming lose their being in losing their becoming, so things whose being is in potentiality exist just so long as their potentiality exists. But, he says, if someone seeks in these cases that which is potential as such, let him assume the moment when the magnitude is not yet being divided and when the contest can be completed but is not yet being completed. Its actuality is being present with the potentiality in the process of bringing to completion, rather than in its having been completed as in the case of the statue. For what is unlimited is not a whole but part by part.

But let the unlimited be potential in the division of magnitudes through their section without limit. But how in general is something divided without limit, and what is it that divides it? For a skill would 494,1 not cut it without limit, for the duration of the life of the craftsman forbids it, and his tools are not capable of it, while if nature were to cut some magnitude without limit it would already have made many segments useless for composition through their smallness. But if it puts them together again before cutting them into the smallest and 5 useless size it will not be unlimited section. So perhaps one should say that nature makes divisions so far as is needed, cutting into two, perhaps very small, parts, and then putting them together again, if it be necessary to cut them again, but does not cut them always at the point of junction but also at another part, and leaves out no individual bit, if it cuts at other parts at other times. And in that way there will appear to be actual cutting without limit, and perhaps no 10 absurdity results.³¹⁵

206a25 Being without limit is different in the case of time [and men and in the case of the division of magnitudes. For in general the unlimited consists in one thing being continually taken after another and what is taken always being limited, but always one thing after another. Also 'being' has many senses, so one should not take the unlimited as a particular thing like a man or a house, but as existing as a day is said to do so, and a contest, whose being is not as some substance, but always is in coming to be and ceasing to be; limited, but always one thing after another. But in the case of magnitudes the part taken remains, while in the case of time and men they cease to be] in such a way as never to be exhausted.

He has shown that being unlimited consists in continuing without limit in the case of the division of magnitudes, the increase of 15 numbers and the extension of time, and previously also in inexhaustible coming to be. So, he says, it is common to all to have their being in coming to be and therefore not to be all at once but in part, and the parts taken one after another are always themselves limited. But he says that it is different in the cases of the division of magnitudes and of time, because in the case of time the part taken on each 20 occasion ceases to be, since the past always has ceased to be; but in the case of division the part taken remains. But neither is time exhausted for coming to be nor magnitude for division. Having said that the explanation³¹⁶ of the existence of the unlimited was one thing being continually taken after another and not all at once, and that for that reason there is process without limit, he adds the other explanation as well; for if the existence of everything were in their 25 instantaneous occurrence there would have been nothing unlimited

in coming to be. But since some things that have their being in becoming exist part by part, like a day and a contest, the given part is always limited and one after another. Also, since the parts are inexhaustible they bring about a process without limit, different, as
30 stated, in the cases of the division of a magnitude and the extension of things coming and ceasing to be, including time and the never ending succession of men, since the cut-off bits of the magnitude remain, but not in the other cases. But even if they remain, division without limit has its being in becoming and it does not exist instantaneously. But those parts that do not remain have their unlimited
35 process not in coming to be alone but also in ceasing to be. He
495,1 contrasted this that has its being in becoming with being which comes to be as substance, like a man and a house. The substance that has come to be exists all at once and not continually in part as in the case of processes without limit, where every given bit is limited but there
5 is the limitless because always another part can be added.

It seems that somehow the same things are repeated after 'being has many senses', the same things being adduced and mostly in the same words. But if, as many copies have it, and in a reading known to Alexander, after 'what is taken always being limited' 'but always
10 one thing after another' is joined on – not that which immediately follows but the one further below –[317] all that comes in between having been removed, the argument would be without repetition and clear. But, if 'also "being" has many senses' etc. is added, perhaps 'also' does not indicate another argument but makes the exposition of the same
15 thing clearer; this is not the custom of Aristotle, who values succinctness,[318] unless perhaps after adding the explanation of the unlimited he clarifies[319] what follows by adding 'have its being in becoming'.

206b3 But that by addition is in a way the same [as that by division. For in a limited thing addition comes about in the reverse direction; for as the thing is seen as being divided without limit it will appear in the same way as being added to a determinate part. For if in a limited magnitude one has taken a determinate part and takes another in the same ratio, not including the same magnitude of the whole, he will not exhaust the limited magnitude. But if he increases the ratio in such a way as always to remove the same magnitude he will exhaust it, since every limited magnitude is used up by any determinate amount whatsoever being removed. So the unlimited does not exist otherwise, but exists] potentially and by exhaustion.

20 He has said earlier that the unlimited exists by addition and also by division, and has also stated the way in which the unlimited in division occurs. Now he says that also the unlimited by addition is in

a way the same as that by division, but in the reverse direction. For as every given limited magnitude is divided, in the same way also to every given undivided and limited one the cut off portions of the divided one are added. It is added to in the same proportion as the other is divided. 'In reverse' can be used in this way as well, because addition is the reverse of division, and can also mean that the addition is with respect to the other part which is not cut off. Still both division and addition are unlimited in similar ways, and from the same materials, but one is by putting together, the other by division. But he himself will also state more clearly the difference between the two in what follows. Having said how both division and addition are without limit, he well adds that not every division and addition is without limit. For if one takes equal cut off parts, e.g. a finger's length each time, one will use up and exhaust the magnitude, and it will no longer be unlimited, because all the limited magnitude is expended, even if measured out by the least determinate quantity. But if one takes cut-off and added bits not of a definite size, but in a ratio, such as a half, or third, or similar part of what is continually left, the division will not end, and for that reason nor will addition. For that is the idea of what is being said.

With regard to the text that says 'for if in a limited magnitude', which seems to contain the explicative connective pointlessly, the idea is like this: having said that division and addition are without limit he adduces as an explanation of this that the division occurs in a definite ratio and not of a definite magnitude, as if he had said 'for if one makes the division in this way, and not in that way, he will divide it without limit'. If 'in the same ratio, not including the same-sized magnitude of the whole' be so written, it makes the idea clear. For if the ratio be the same in each of the cuts, e.g. the half or a third of what is taken on each occasion, but it is not the same magnitude from each whole put forward for division on each occasion that is taken away, e.g. a finger's length, it will not exhaust what is limited. But if the reading be, as in some copies, 'not including the same portion of the ratio' the statement is less clear, but might mean the same as what was said previously.[320] For 'not the same term of the ratio' does not mean 'the ratio being of terms the same as of the whole', but the equivalent of 'not including the same definite magnitude in what is taken away from the terms of the ratio', which was also the term of the ratio available for division at the beginning, e.g. if the magnitude being divided was three yards and it was divided by a ratio of a third, but the bit cut off was two yards, one should divide it by the same ratio of a third, not however by the same portion of the ratio as in the previous cut, e.g. a yard, (for in that way the whole will be used up) but in the same ratio of what is taken on each occasion, not the same magnitude as in the beginning.

But why should the division be in the same ratio? What prevents taking a half of the magnitude at the beginning and a third of the half? For so long as some magnitude is left it is divisible and the division does not come to a halt. Or was 'in the same ratio' not said idly? For that becomes the cause of the same magnitude not being taken away on each occasion. For the third part of both the greater and the lesser magnitude, even if they are in the same ratio, still are not the same; but if the ratio of the division is continually different nothing prevents the same magnitude being taken from the greater and the lesser. For if we take the sixth part of what was six fingers long and the fifth of what remains and again a quarter of what remains, and so on, thus taking away a finger by each division, we shall use up the whole and the division will not be without limit. That is therefore the reason why the division ought always to be in the same ratio. For in that way the same magnitude will never be taken, so as to exhaust it. Also this is what he made clear by 'but if one increases the ratio so that the same magnitude is always included, one will exhaust it'. For, as was said, the fifth is greater than the sixth, the quarter than the fifth, and so on, since the same portion, a finger, is taken on each occasion. 'By exhaustion' was said as equivalent to 'by division' and 'by decrease'. But he was content with 'by exhaustion' since because of this addition also is without limit.

206b13 Also it is in actuality[, in the way that we say that a day is, and a contest. Also it is potential in the way that matter is, and not in itself as is the limited. Also that by addition is potentially unlimited in much the same way as that by division. For it always will be possible to take something extra, but it will not be beyond every magnitude, as in the case of division it is beyond every definite magnitude and will always be less. But it cannot exceed every magnitude by addition even potentially, since it is not in fact actually unlimited, as the naturalists say the body outside the universe is unlimited, whose essence is either air or something similar. But if it is not possible for a perceptible body to be actually unlimited in this way, it is plain to see that it would not be so even potentially by addition,] except as it has been said to be so in reciprocity with division.

He has said how the unlimited exists in the division of magnitudes, that it is by being able to be divided, and in general potentially. Now he says that these exist in actuality as well, not as things which are present all at once, but as those whose being is in becoming and are present by part, in the way that a contest and a day exist. For these also have their potentiality mixed with actuality. Their actuality exists just so long as when they have their potentiality, as was said in the case of change. He makes clear of what sort this inherent potentiality is through the illustration of matter. For as matter, which is potential and present in composite and generated things, as it participates in a different form at different times is actual insofar as it is assumed, but, as potential being of a nature to take different forms at different times, brings about inexhaustible coming to be by being of a nature to take different forms at different times, so, too, in division the given magnitude is always actual, but, because every given one can always be divided, its division is retained as inexhaustible. And as matter does not have a shape of its own, but only by partaking in a form, so, too, potentiality in division is not limited in its own nature, but it is unlimited in that respect and is the cause of a magnitude's being able to be divided without limit; but insofar as each part taken is actual it is in that respect limited. But it is not only in regard to division that potentiality is the cause of being unlimited, but also in with regard to addition. For by actually adding we shall never attain to the unlimited, but what is taken in actuality is always limited. But being unlimited comes about in addition by it always being possible to add something to this from the portions that come to be without limit.

He has said that we count the unlimited by addition as being in a way the same as that by division, and that its always being possible to take something extra is common to both. For, in the case of division, division without limit is possible because of the fact that there is always something to divide, and, in the case of addition, because there is always something to add, which came to be by continual division, and this is the something extra which it is possible to divide or add that is common to both. But they differ from each other because in the case of the unlimited by division it is possible to get a lesser magnitude than every given magnitude, but in the case of that by addition it is not possible to find a greater than every magnitude. For there is not something extra beyond every magnitude, since the universe is limited and nothing is beyond it, as was shown in the first book of *Cael.*;[321] but there is something extra beyond that magnitude which, by being cut on some sides, was increased by the sections cut off from the other. However, it is not possible even potentially to exceed continually every given magnitude by addition, as was the case with other addition and division. For it would then be possible then when there was some body which happened to be unlimited, because there was some something extra beyond every given limited magnitude from which it would be possible to add to it, as some of the naturalists say that there is. But he reasonably recalls those who say that the actually unlimited is among attributes, and not those who say it is substance, either because that through addition and division has to be a quantity, which happens to be unlimited or limited, or because the discussion is about the natural. But the natural that was unlimited, if it existed, would be a body which happened to be unlimited. Or perhaps it is because the unlimited as substance does not gain its unlimited character by addition, but has it from its nature, just as each of its parts, if it were to have them, would be unlimited. So if it was shown previously that there is no body which happens to be unlimited, it also is not possible to get a magnitude greater than every given magnitude; but process without limit is possible in the case of addition, as was said earlier,[322] only by adding the portions from one segment to the other, which would be reciprocally.

But someone might raise the problem why, if there is no unlimited perceptible body in actuality, there also could not be potentially addition without limit. For see, there is no actually unlimited number, and yet in the case of numbers there is addition without limit. This suggestion will be examined again in what follows.

206b27 Since Plato also [made the unlimiteds two for this reason, that there seems to be excess and process without limit both in increase and in exhaustion. But having made them two he makes no use of it; for in the case of numbers there is neither unlimited process in exhaustion since one is the least, nor in increase,] for he limits number to ten.

It is of considerable concern to Aristotle to exhibit, if possible, his accounts as consonant with those of famous men also at the superficial level, and, if that is not possible, at least to render them as in agreement with the ideas of the ancients. At least he says that Plato was moved by the same ideas at one time to make the unlimited the indefinite dyad, hinting at the indefinite process of division and composition without limit, at another time to say that it was the great and small, indicating the unlimited division leading to the lesser by the 'small', the process to the greater through addition by the 'great'. But having stated the community of ideas he also next adds the lack of fit of the apparent account of Plato's elsewhere with the idea now mentioned. For Plato places the unlimited among principles and says that numbers are principles, since the ideas were the principles of what exists and he said that the ideas were numbers, but he was unable to see in them the unlimited by division, for the unit is indivisible and division is limited by it, nor that by addition. For he held the increase of numbers to be as far as ten, like the Pythagoreans who said that ten was the complete number and the whole number.[323] For after ten there comes next a recirculation of the same numbers and, as it were, a bending back, which happens to no number within the decad. And it is not so with respect to any numbers whatsoever beyond it, but with respect to the decad. For 20, 30, 100, 1,000 also are a composite of such decads.[324] However, the unlimited process to exhaustion occurs in the case of magnitudes, but that to increase and addition does not occur simply in the case of magnitudes. For it is not possible to find a greater in the case of every given magnitude; rather addition without limit proceeds only in that way which adds the segments from unlimited division. But, if at all, it is in the case of numbers that there is growth without limit, because it is possible to find a greater than every given number. But if number proceeds only as far as ten it is not possible to obtain unlimited addition even in the case of number.[325]

Thus at the superficial level the unlimited seems to be rejected by those who hold that numbers are principles and take them as far as ten. But if those who say that numbers are the principles of things placed the unit as a limit and the indefinite dyad as the unlimited, and these as principles, it is clear that they did not reject the unlimited. In general, if they denied increase of number beyond ten, perhaps someone could say that they did not leave unlimited addition in number; but if they posit secondary and tertiary[326] units, and recognize combinations of decads, (and Plato enumerates thousand-, three thousand- and ten thousand-year cycles of souls, and honours the twelve gods), from this it is clear that in setting the number of principal numbers as ten he does not prevent the addition of number from increasing without limit.

206b33 So it comes about that the unlimited is the opposite [of what they say that it is. For it is not that beyond which there is nothing, but that beyond which there is always something which is unlimited. Here is a sign of this: for they also say that rings that have no bezel are unlimited because one can always get further. They say this through some similarity, but not strictly; for this has to be so and also the same place as before must never be reached. In the case of a circle that is not what happens, but always only the next bit is different. So that is unlimited of which, taken as a quantity, it is always possible to find a part beyond. That of which nothing is beyond is what is complete and whole; for that is how we define a whole, as that from which nothing is missing, such as a whole man or a cup. As the particular, so the strictly whole is that beyond which there is nothing. That of which something further is missing is not all whatever be missing. 'Whole' and 'complete' mean altogether the same, or the very similar. But nothing is complete (*teleion*) that has not a terminal point (*telos*), and the terminal point is a limit. Therefore one should believe that Parmenides spoke better than Melissus. For Melissus says that the whole is unlimited, Parmenides that the whole is limited, 'equal in all directions from the centre'.] To connect the unlimited with everything and the whole is not to join flax to flax.

He has indicated the nature of the unlimited as it occurs in division and addition and as seen with regard to process without limit. Next he proposes to correct incorrect ideas about the unlimited. For if we have well located the unlimited it turns out that the unlimited is the opposite of what some people propose. For they say that the unlimited is that beyond which it is not possible to find anything, but we say that it is that beyond which it is always possible to find something. For in division there must always be something to be divided, and in addition something to be added. And let nobody think that we are saying something incongruous when we say that the unlimited is that beyond which it is always possible to find something. For such a use of the word is common, since they call rings that have no bezel unlimited, because it is always possible to find a bit beyond what one has got to. And this circular ring or a circle, as he will later call it,[327] has something similar to the unlimited; but it is different insofar as in the case of the circle, even if we can always find a bit beyond, still it is the same bit again as we go round it, but in the case of the unlimited it is always another and never the same. It is as if such a limitlessness were in a straight line and not cyclical, as in the case of circles, whose rotations can be counted since they repeatedly start at the same point and are not strictly limitless, but only through some similarity, since there is always something beyond in each case. But the unlimited must also never reach the same point, so as a good definition we say not that beyond which nothing can be found but that of which, taken as a quantity, it is always possible to find a part beyond. It is clear that in saying 'taken as a quantity' he included the unlimited as continuous and as divisible; for each is a quantity. But perhaps by 'taken as a quantity' he was indicating the difference between the strictly unlimited and circles. For in the case of the circle when, having gone round it, we start from the same place again, there is nothing quantitative beyond, since nothing is added to the circle; but in the case of the strictly unlimited there is always something beyond the given quantity, either towards the greater, as in the case of addition, or towards the lesser, as in the case of division; and there is always division and addition beyond the given quantity. Also 'quantitative' is appropriately added in another way, because the addition is not of a form but of a quantity; for the addition is not made as to a body but to a body as a certain quantity. But Alexander explains 'as a quantity' in another way as well. 'For', he says, 'what we are investigating, and of which we say it is unlimited, is that which is in the category of quantity; but if it were in that of quality we would say "that of which, taken qualitatively, there is always something beyond".' But perhaps quality as such does not participate in the difference between the limited and unlimited; but, insofar as quality participates in quantity, it also was said to be as such limited or unlimited in some way.

Having shown that it is more suitable to say that the unlimited is that of which it is always possible to find something beyond, he next shows that what the others said, that it was that beyond which there was nothing, is not suited to the unlimited but rather to the limited, arguing as could be put in this way: that beyond which there is nothing is a whole; for a whole is that of which no part is missing; but the whole and the complete are either the same or nearly so, for the complete is what has an end and what has an end has a limit; but what has a limit is limited. Also there is the demonstration from the definition which the others take as being of the unlimited, but which ordinary use recognizes as being of the whole and total. He reminded us that the whole is that from which nothing is missing from particular cases, natural and manufactured, such as a man and a cup, and concluded that, as in particular cases that thing is whole of which no part is missing, so in the case of the strictly and unqualified whole. But that is strictly whole which is the whole of being. For each partial whole is not strictly a whole, because a part is of that which is a whole, and the latter is, the former is not, a whole. The whole of being is strictly a whole because it is not a part of anything and contains everything within itself. Perhaps 'strictly' can be taken as 'universal' as well in his usage. For it is rather to this that the particular is contrasted. Whichever way, it is worth noticing that for him particulars do not strictly exist. So that strictly exists from which nothing is missing.

He well said that the whole and the complete are either altogether the same or closely akin in their nature. For they are the same in substrate, but in account they are different but very similar to each other. For the complete is what has a beginning, a middle, and an end, while the whole is viewed as what has a fusion of these. For a whole is something continuous. So when, for example something full is discrete we strictly say 'all of it' and not 'the whole'. Also he supposes that the complete has some distinction between the beginning, the middle and the end. But the whole comes about as something uniform and containing its parts together, so as not to be completely unified, for then what passed into each other would not be parts any longer but, if anything, elements. But nor are these completely discriminated, for then they would be kinds and not parts, but would rather have their being in being discriminated and exist as discriminated. So a whole consists altogether in what partake of it and being complete is prior. For it is not possible for something to become whole if it is not first fully supplied with beginning, middle and end. So, as one would expect, participating in completeness comes from being whole.[328]

If what is whole and complete is limited, Parmenides' account of being is better than that of Melissus. For Melissus, having said that what exists is unlimited, also says that it is a whole;[329] but Parmenides says that it is a whole, as 'equal in all directions from the middle' makes clear, and, since he says that it is a whole, reasonably says that it is limited;[330] for what has a middle and extends equally all round in all directions inevitably also has an extreme. But Melissus combines opposites. For if the unlimited is a whole and a whole is limited the unlimited would be limited. So he who says that the unlimited is whole and everything weaves together things not of a nature to be woven together. For the proverb refers to doing this.[331]

207a18 For it is from this that they derive the dignity [ascribed to the unlimited, that it encompasses all things and contains the universe within it, since it has a certain resemblance to a whole. For the unlimited is the matter of completeness of magnitude and the potentially, but not in actuality, whole. It is divisible towards exhaustion and the reciprocal addition, but it is a whole and limited not in itself but through something else, and *qua* unlimited it does not contain but is contained. Therefore it is also unknowable *qua* unlimited; for matter has no form. So it is plain that in account the unlimited is rather a part than a whole. For matter is a part of a whole, like the bronze of a bronze statue, since, if it contains among perceptible things, among things intelligible the great and the small should contain things intelligible. But it is absurd and impossible for the unknowable] and indefinite to contain and to limit.

Having said that the description 'beyond which there is nothing' fits a whole, he says that for those who transfer it to the unlimited the cause of their dignifying the unlimited as containing all things is that they think of it as a whole because they transfer the definition of a whole to it. And perhaps there is some excuse for the fallacy. For the unlimited has some similarity to the whole. But if it is not a whole it also will not have the pretensions of a whole. How would the unlimited be a whole, since a whole is that beyond which there is nothing, but the unlimited is that of which it is always possible to find something beyond? But the unlimited seems to be like a whole because, he says, the unlimited is the matter of the completeness of a magnitude. For the whole and complete magnitude is a composite of matter and form and contains the unlimited through its matter, but it is limited through its form. But because matter is potentially what the composite is, and the potential has a certain similarity to the actual, for that reason the unlimited is said to be in a way similar to the whole, which is the composite. And thus they next predicate of the unlimited the characteristics of the whole, which are to contain all things and to have all things within it. But matter is unlimited, both because it is by its definition indefinite and unbounded, while form provides a limit and definition, and because the cutting without limit of a magnitude and its increase without limit come about through its matter, while its form remains constant. But if form is always something beyond its matter, matter is also in that way unlimited in itself, but through its form it is able to be limited and a whole. So the unlimited does not contain, since it is viewed as matter, but rather it is contained. For what limits contains the limited and form contains matter. But if all knowledge, which is a sort of boundary and grasp, is through form, while matter through its own account neither is form nor has form, while matter is the unlimited, the unlimited is reasonably said to be unknowable. For it was said also in the discussion of matter that matter is barely knowable by analogy.[332] But if the unlimited is viewed as being matter, and a whole as a composite, and matter is a part of the composite and whole, as the bronze is of the brazen statue, it is clear that the unlimited must have the status of a part since it is in a whole.

He has shown that the unlimited is contained rather than containing and that it is unknowable by its own nature and now examines the superficial interpretation of Plato's discussion. For Plato said in his lecture *On the Good* that matter was the great and the small, which he said was also unlimited, and that all perceptible things were contained by the unlimited and were unknowable because their nature was enmattered, unlimited and fluid.[333] Aristotle says that it seems to follow from such an account that the great and small THERE, which is the indefinite dyad and itself a principle together with the unit of all number and all things that there are, is present among intelligible things.[334] For the forms are also numbers. So it follows also among intelligible things that the intelligibles, which by their nature are knowable and definite, being forms, are contained and bounded by the unlimited and unknowable THERE. Now this is, superficially, the absurdity of the account. But one should consider that he reasonably says that enmattered things are contained by matter and material limitlessness not as bounding them but as pervading them all and, as it were, giving them form, and reasonably material limitlessness makes things HERE unknowable. However, intelligible things are immaterial and pure forms and contained by the immaterial limitlessness THERE which brings about the distinction of the forms through the dyad and provides the more and less by excess and deficiency through the order THERE and the inexhaustibility of potentiality.³³⁵ They would not be unknowable through that sort of containment. For also the limitless HERE, being material, made things HERE unknowable, but that other, being controlled by the one and by limit, is the wealth and fruitfulness of the forms. But if the effusion of the forms THERE into being and intelligible unification comes about through the limitlessness THERE, it is not surprising that the forms should spill over through this effusion and their knowable nature. For what is knowable tends to limit, but the unknowable and ineffable to limitlessness.

Chapter 7 begins here in modern editions.

207a33 Accordingly it results also that it seems that there is no unlimited magnitude by addition, [so as to exceed every magnitude, but that there is by division. For matter and the unlimited are contained within,] and form contains them.

Having shown that the unlimited does not contain but is contained, he says that in the case of magnitudes the reason for this dissimilarity in the unlimited is that there is no process without limit by addition such as to exceed every magnitude, but that by division is such as to be less than every magnitude. The cause of this is that matter is contained within as being unlimited and the cause of limitlessness, while form contains it. Therefore, for those who wish to make a magnitude smaller without limit, matter, which is present through its limitlessness, provides them with the means. Also it is through matter that addition without limit from the bits cut off comes about; but it is impossible for the boundary to fall outside the form, which is a limit. Or otherwise it will come about that the form will also be increased; but a form does not become greater nor lesser. For, if there were an unlimited by addition such as to exceed everything, there would be limitlessness with respect to the container which delimits and bounds; for it is with respect to this that addition occurs. But what bounds is form. Also in that case there would be limitlessness with respect to form. But this is absurd, for form is the cause of limit and not of limitlessness. But when the absence of limit is in division there is no absurdity. For matter which is the unlimited is contained, and division and addition through division³³⁶ occur in what is contained. But it will seem that matter will create magnitude through its own character, if division is through it. Or perhaps one should not say that matter is divided, but that a magnitude is divided with respect to its matter, which is the unlimited within it.

207b1 It is reasonable for one[337] to be for number [a limit in process towards the minimum, while in increase it may exceed every amount, but that in the case of magnitudes, on the contrary, in decrease they may exceed every magnitude but in increase there should not be an unlimited magnitude. The reason is that the unit is indivisible, whatever the one may be, e.g. a man is one man and not many, and a number is a plurality of units, and just so many, so that it must come to a halt at the indivisible. For three and two are paronymous terms, as is each of the other numbers. But it is always possible to think of an increase. For there is an unlimited number of divisions of a magnitude. So it is potential and not actual; but what is taken always exceeds every definite amount. But this number is not separate,[338] nor does the limitlessness persist, but it is always coming to be,] like time and the number of time.

He has stated the explanation of the difference with regard to limitlessness in the case of magnitude, through which decrease below any given magnitude is always possible, but increase above is not, and now he sets out the explanation of the difference in limitlessness between number and magnitude, why it is that in number it is not possible to decrease without limit, since we finish at a limit which is the unit, but it is possible to increase, since it is possible to increase every proposed number; but in the case of magnitudes, to the contrary, it is possible in decrease to exceed every magnitude, but not to do so in increase. He states a first explanation why the division of number does not proceed without limit: this is that a number is composed of many units, and what consists of units can also be divided so far as one, but what is one is as such indivisible. So how can it be possible for division to continue without limit? For we come to a halt when we have reached units that are indivisible. Also it has already been said, and is obvious, that a thing is finally divided into the first elements of which it is composed. He reminded us that a unit is indivisible by 'e.g. a man is one man and not many'. For, as a man, he is indivisible; for he would divide into men. For the division into head, hands and feet is not of him as man but as a composite of parts. For being one is a form and so definite, and what is definite is not divisible in respect of its definition. It is irrelevant that each number, being one, is indivisible in respect of its own form, but is divided in respect of the units it contains that have the status of matter within it. But this unit or monad has no matter and is nothing other than an indivisible form.[339] In general, if a unit were divided what would it be divided into? Perhaps always into halves; but each of these is also a unit, no less than the original one. For a unit was to be the minimum in number. So how could it be divided? He reminded us that a number consists of units by 'a number is a plurality of units and just so many; for "three" and "two"³⁴⁰ are paronymous words'.³⁴¹ So if a number is nothing other than a plurality of units and just so many, and each number is named from the number of units in it (for two is from two units, three from three), it is clear that a number is composed of these. So if a unit is indivisible and a number consists of units, and if it is also divisible into the things of which it is composed, it must halt at the indivisible, which is the minimum and the unit.

Having stated the explanation of the halt at the indivisible he next adds the explanation of the advance of numbers without limit in increase, saying that this advance without limit comes about from that in magnitudes; for the increase of numbers without limit comes about also from the fact that magnitudes can be cut without limit because of the limitlessness of the matter contained in them. In both cases the absence of limit derives from matter. That is why he himself added 'but this number is not separate'. Rather, it is clearly enmattered, since in this case also the limitlessness comes about through matter. But the addition which comes about from the unlimited cut-off segments of magnitude is unlimited increase in magnitude, not in number. So if we were to take the unitary number without supposing magnitude, is it not necessary that in this case also it be possible to add a unit to every number taken? In general, from what division comes the continual addition in number in the case of men and other things that come to be? From what concomitant division will be the increase of the number of heavenly change and of time? So perhaps he does not say that the unlimited increase of all number comes about from the division of magnitude, but of that which is in existence and persists. For how could what is not existent nor persists increase or receive some addition? So we should not say that the monadic number is increased, since this has its being only in conception, nor that of men or change or time, or in general of things that come to be and cease to be. For their numbers have their being in the ceasing of the earlier. The earlier ones do not persist in such a way that those that come next can be added to them and the initial number be increased, except in our imagination, which itself does not persist. But this increase could only occur in the case of the addition of cut-off segments of magnitudes, sometimes regarded as magnitudes, sometimes as units, and thus sometimes being increased without limit as a magnitude, sometimes as a number. This means that the actual never achieves limitlessness, whether it be a magnitude or a number. But the capacity to proceed continually without limit is present potentially, because of the potentiality and limitlessness of the matter present in them; and it is because of the matter that process without limit occurs also in the addition of enmattered numbers. Aristotle also appears to indicate the solution that I have suggested by saying: 'the

Translation   131

given continually exceed every definite amount. But this number is
not separate. But nor does this limitlessness persist, but it is always   20
coming to be like time and the number of time'. For he says that even
if the given always exceed every definite amount, still this is not to
be viewed as in the case of some separate number that is conceptual-
ized, but of the embodied number, in the case of which the division
which is the cause of addition for persistent things is without limit   25
through their matter. For the limitlessness is not static, but, as in the
case of change and time and the number that measures them, the
limitless does not persist but is continually coming to be, with the
exception that in the case of addition from the segments cut off the
previous persists and receives what follows, even if the limitlessness
does not persist in this case either, but has its being in becoming, as   30
in the case of change and time and the unitary number. But these
cases have a different sort of coming to be of limitlessness, as the first
ceases to be and receives no addition. But what for them also is the
cause of the process without limit, as the enmattered segments of the
magnitudes cut off are for those that persist? Perhaps it is the
continually present potential, which is itself material, that provides
the continual capacity, the number of time being the prior and   35
posterior, which is given without limit by becoming one after another?
He calls the number of time that through which time both counts and   507,1
is counted.[342]

207b15 The opposite holds in the case of magnitudes; [for the
continuous is divisible into unlimited bits, but it is not unlimited
in increase. For the quantity that may possibly exist also may
actually exist. So that since there is no unlimited perceptible
magnitude there cannot be an excess over every definite magnitude;] for it would be greater than the universe.

He earlier stated as an explanation why there is limitless division in   5
the case of magnitudes, but not limitless addition – that since matter
is the unlimited it also provides an internally contained field for an
unlimited process in decrease, while, since form is a limit and contains externally, it does not permit unlimited addition. But now he
gives the explanation of the continually unlimited process in decrease
– that the continuous is a composition of continuous parts and a   10
magnitude one of magnitudes, while everything is divisible into that
of which it is composed. For a magnitude has parts that are magnitudes and persists through cutting and division into these, as it does
through analysis into elements. It is common to both cases that a
thing is divisible into the things of which it is composed. If it is being
cut, the cut is always into parts, if it is being analysed, the analysis   15
comes about into elements. He will show in Book 6 that a magnitude

consists of magnitudes from the fact that it is of things either with parts or without them, while things without parts when put together do not bring about extension.[343] He says that the explanation of a magnitude's not being increased without limit is that as we proceed to the greater we meet with the whole and the universe, which is bounded, just as when we proceed to the lesser in numbers the unit which is bounded meets us and halts the process towards the unlimited. But if it were possible to add to every body, there would be no universe or whole, but something such as the naturalists say is an actually unlimited perceptible body, which has been shown not to exist. But if every magnitude were capable of increase without limit, so that the continually greater potentially existed, it would be a certain quantity in actuality. 'For the quantity that may possibly exist may also actually exist.' So since there is no actually unlimited perceptible magnitude, as has been shown, so it is not possible as we proceed to exceed every definite magnitude without limit so that nothing may actually be greater than the cosmos, which is the whole and total magnitude, not only according to Plato but also to Aristotle.[344] They both call the whole cosmos the heavens, either naming the whole from its most important and largest part, or because to be turned towards one's own creator and look upwards (from which also heaven is named)[345] holds not of the everlasting[346] body alone but of the whole universe. He will show particularly in *De Caelo* that there is no magnitude greater than the cosmos, nor anything outside the heaven,[347] but it has, I think, already been shown by the arguments that refute the unlimited magnitude. For if there were something still greater than the cosmos that remained indefinite it would be unlimited.

But what does he mean by 'for the quantity that may possibly exist may also actually exist'? For it is not the case that since it is possible to add to every number there is also actually an unlimited number, nor that since there is not actually an unlimited number the unlimited addition of number is already for this reason prevented. For it is possible to add to every given number. So someone might say: even if there is not actually any unlimited what prevents the possibility of increase without limit being potential? For the potentiality in division is through the possibility of getting something smaller, without it being the case that sometime the least will be reached. So what prevents the potentiality in addition being thus, so that it is possible always to get something greater, but never to reach the maximum? But, if that is the case, one should not deny the existence of addition without limit because there is no unlimited perceptible magnitude in actuality. Well, perhaps, as in the case of magnitudes that is why it is not possible to exceed every given magnitude – because there is not actually an unlimited perceptible magnitude – so in the case of numbers. For in their case also it is not possible to exceed every given number except conceptually, because there is not actually any unlimited number. But it will soon be shown that conception often is of the unreal. So what on this view did we mean when we said that number was affected in the opposite way to magnitude, insofar as number always is without limit in increase, but magnitude in decrease? Perhaps that does not hold of every number in existence, but only of that from dichotomy, while magnitude increases without limit through the process of dichotomy without limit. So as it is impossible to exceed every magnitude, since there is no unlimited perceptible magnitude, but it is possible to add magnitude to magnitude without limit by adding without limit the fragments cut off by unlimited dichotomy, so it is impossible to exceed every number except conceptually, since there is no actually unlimited number. It is possible to add always only to a number resulting from continual dichotomy through the absence of limit inherent in matter, not doing this in the case of separate number, but of that which proceeds together with dichotomy.[348] Perhaps if we say this we shall seem to make sense. But perhaps we do not, because even if magnitude allows unlimited addition through division, that which receives the addition still does not exceed every magnitude. For it has the limitlessness resulting from division, and that limitlessness is from inside and does not proceed beyond. But why does not number in dichotomy without limit, which itself proceeds without limit, exceed every number, even if there is no actually unlimited number? So once again our problem[349] comes back to the same point: what does he mean when he says that for this reason there cannot be an excess over every definite magnitude – that there is no actually unlimited perceptible magnitude – since, even if there is no actually unlimited number, it is necessary that the number resulting from unlimited dichotomy should exceed every number? Well, perhaps it was impossible in the case of magnitude for there to be an excess over every definite magnitude unless the magnitude was unlimited; for everything definite is limited. So, if it were possible to find a greater than the greatest of everything definite, that found would have to be indefinite and unlimited. But in the case of the addition of number it is necessary, even if there is no unlimited number, that addition by dichotomy should continue without limit and be greater than every given number. For also dichotomy without limit occurs without ever reaching a minimum in the way that the increase of magnitudes reaches a maximum which is the whole and the universe. So why is it not true also in this case that every limited number is definite? So if we should wish to find a greater than the greatest definite and limited number, it is necessary to spill over into the unlimited. For the unlimited number is greater than the greatest definite and limited number. So there also cannot

be an excess over every definite number unless there is an unlimited
number. How, then, is there unlimited addition of number and how
unlimited division of magnitude, if there is no minimum magnitude,
if what was said is true – that the quantity that is possible potentially
is also possible actually?

207b21 But the unlimited is not the same [in the cases of
magnitude, change and time, so as to be of a single nature, but
the later is used in relation to the prior; e.g. change is so because
the magnitude with regard to which it moves or alters or
increases is so, and time because of change. We now make use
of them, but later we will say what each is] and that[350]
every magnitude is divisible into magnitudes.

He has said that the unlimited occurs in many contexts (for it is found
in magnitude, in number, in change, in time, and in all processes
without limit). He now well notes that the unlimited which occurs in
the continua of magnitude, change and time does not do so as the
same and a single nature and single kind in all the cases mentioned.
For it is not predicated of them on an equal footing, as is proper within
kinds, but of one as prior, the other posterior.[351] For since being
unlimited holds of these as being continuous, because the continuous
is what is divisible without limit, and magnitude is what is primarily
continuous, this also is primarily divided into magnitudes since it
consists of magnitudes, and for this reason it is also divisible without
limit. Change is continuous because of magnitude; for it too is continuous and divisible without limit by occurring in a continuum,
magnitude. Time is so because of change; for it is of change. For time
is the number of change.[352] So being unlimited is not predicated in
the same way or as a genus, but always of the second through the
first. For this reason it has a prior and posterior. Predication as of
prior and posterior occurs also in the case of similarly named terms,
for perhaps a grandson may have the same name as his grandfather
in this way; but it also occurs among terms named from one case.[353]
The predication of limitlessness is rather of that sort. He now reminded us briefly that change is continuous derivatively from magnitude by saying: 'change because the magnitude with regard to
which it moves or alters or increases is so'. He takes 'with regard to
which it moves' as equivalent to 'change of place'[354] and reminds us
through formal changes as well that all kinds of change are of
magnitudes and in magnitudes, and that they are continuous because
of the magnitudes. But now, as I said, he reminded us briefly of this,
and, going on, that magnitudes are the primary continua and divide
into magnitudes, since they consist of magnitudes. This is why it first
of all exhibits division without limit, and change because of magni-

tude, and time because of change. The definition of change has been stated, and those of time and magnitude will be stated.

207b27 This account does not take away [their study from mathematicians when it denies that there is an unlimited in such a way as to be actually inexhaustible in increase. For, as it is, they do not need the unlimited, since they make no use of it, but only that a finite line shall be as long as they please. But any magnitude whatsoever may be divided in the same ratio as the greatest magnitude. So for their proofs] it will make no difference whether the infinite occurs in existing magnitudes.

A man of understanding will not destroy anything prized by the wise through his statements. So geometry is justly prized and geometers prove their theorems in it as if there is an unlimited magnitude inexhaustible in increase. For even in their postulates they assume that one may produce a limited straight line continuously in a straight line, and, if that is so, that it is possible to exceed every given line, and, if that is so, that it is necessary for there actually to be an unlimited magnitude. Thus the foregoing arguments are refuted in several ways. But often in their proofs as well they suppose unlimited lines, so that it will seem to many that it is necessary either that there is an actual unlimited magnitude or that geometry is refuted. Also it is clear that a wise man would agree to anything rather than to the refutation of geometry. For in addition to other merits it has this supreme one, as Plato says, 'of aiding all studies, so that they will be better acquired'[355] by one who has studied geometry than by one who has not. To assuage this fear he says that the argument refuting the actual unlimited does not take away their study from mathematicians. For mathematicians nowhere make use of an actual unlimited magnitude, but rather complete all their demonstrations with limited lines and planes. So they have no need of the actually unlimited magnitude, but in order to have as great a one as they want they suppose unlimited lines. So even if they postulate the extension without limit of a straight line in a straight line, in order not to be embarrassed by there being no longer a straight line from which to remove a part or to join to or to extend, still the joining or cut falls in one or another part of the plane surface, and they always use a limited straight line. So if mathematics does not need an unlimited line he who abolishes it does not abolish any mathematics. He not only showed inductively that they have no need of the unlimited for cuts and joining by reminding us that everywhere they use limited magnitudes, but he also demonstrates it geometrically. For if it has been demonstrated that if one takes a given uncut straight line, even if very short, it is possible to cut it in the same ratio as one already cut, even the very longest, it is not necessary to produce a proof in the case of that which it is no longer possible to extend, if it makes no difference to the proof which is given. For he who has shown in the case of the very longest straight line what is possible to show in the case of a shorter, and who cannot extend it as far as he needs because it is the longest, is himself to blame for his lack of space and inability to carry out the proof, not he who refutes the actual unlimited. But in their constructions they not merely do not[356] take unlimited lines, but not even the longest. For who makes use of the diameter of the cosmos in his diagrams? He well showed the lack of need of the unlimited by the case of cutting in the same ratio, from which unlimited division is proved by geometers. It is as if he said that even on the basis of their own proofs they have no need of an unlimited straight line that actually exceeds also in increase, but they use being unlimited in the case of division, and this they demonstrate. Both of these are proved by one theorem in Book 6 of the *Elements*, of which the proposition reads: 'to cut a given uncut straight line in the same ratio as a given cut line.'[357] Having said that it will make no difference to the demonstration if one constructs the proof in the case of the shortest line, which can also be extended, he added 'but existence belongs to existing magnitudes'; that is to say, even if it is possible to construct the proof also in the case of not actual but conceived magnitudes, still the existing figures will not include all magnitudes. For they include the actual rather than the conceived ones as well.

Alexander inquired how the first theorem[358] of Euclid's *Elements* is not refuted, if it is not possible to extend a straight line or to draw a circle also beyond the universe. For if the given limited straight line on which it is required to construct the equilateral triangle can be the diameter of the cosmos, it is impossible to construct an equilateral triangle on this if there is nothing outside the cosmos. For the diameter of the universe is through the centre of the circles from the intersection of which the lines joined to the extremes of the given line make with it the equilateral triangle. Having raised this question, he answers it by saying: 'Since that is unlimited of which, taken as a quantity, there is always something to take beyond, as has been proved, it is clear that mathematicians assume that the lines that they assume to be unlimited are such that they can be increased. For those lines are unlimited beyond which there is something. But one cannot increase the diameter of the universe. So they suppose them as less than the diameter, which they assume to be limited, since those to which they can add and which they can extend are unlimited.'[359]

It is worth noting that Aristotle turns that argument in the opposite direction. He has proved that that is unlimited beyond which something can always be found, and introduces this other conception of the unlimited which is true of another unlimited, that of unlimited process. However those who posit an actual unlimited, whether in lines or in other magnitudes, said that that was unlimited beyond which there is nothing to be found. So it is better to say, first, that they postulated such a limited line as that beyond which something[360] more could be found, from which they extend the two lines equal to the first line. So if there is something outside the universe there is nothing to prevent maintaining the theorem also in the case of the diameter of the universe. But, if there is nothing outside, this given line does not exist. Further, if mathematical lines were natural and existing in a place, it would be necessary to postulate some place outside the universe; but if they are conceptual objects their place is the mind of the thinker and not a natural place.[361] For that is why mathematical objects are said to be abstracted, because we take away all the characteristics of natural magnitudes – place, quality, time, affecting and being affected – and examine only the intervals and the quantity as being continuous. So he who takes away from us place outside the universe does not take away the demonstration with it. For we do not draw and prove the diagrams on the basis that the lines are here or here. So just as Democritus, when he tries to establish that colours are not naturally present in bodies, but have their existence by convention and posit,[362] would do no damage to geometry, since one does not in general use colours in demonstrations, so he who abolishes unlimited place and place in general in no way impedes graphic proofs.[363]

For what is taken away is something of which the provers have no need. It will be proved in the next book that mathematical objects are not in place.[364] He who supposes simple and minimal principles and elements hinders geometry;[365] for he abolishes cutting without limit, together with which many things now readily demonstrated in mathematics are abolished.

207b34 Since causes are divided into four, [it is plain that the unlimited is a cause as matter, and that its being is a privation, while the self-sufficient object is that which is continuous and perceptible. All, including the others, can be seen to treat the unlimited as matter; which is why it is also absurd to make it what contains] rather than the contained.

He has shown that there is not an unlimited thing, as the generality believe, but in another way there is, as that which proceeds without limit, and shown also that it is the potential and that beyond which it it is always possible to get something extra. For those who posited the unlimited said that it was a principle and he himself showed that if the unlimited exists at all it must be a principle and not derived from a principle; and since in any case the discussion is now about natural principles and everything here propounded must be referred to them, so reasonably he also shows to which of the natural principles and causes the unlimited is connected. Indeed, he reminds us of the division of causes into four – the material, the formal, the efficient and the final – and says that it is plain that the unlimited is a cause as being matter, by 'plain' bringing the induction before our eyes. For who would call either the end aimed at, which is a limit, unlimited, or the efficient cause which from its own bounds makes thing definite, or form which is manifestly a limit of the compound? So rather it is a cause as matter, not because matter from its own account and as a substrate is the unlimited, but because it was an attribute of it to be privation, which is limitlessness, before receiving form, and because privation has been shown to be something else prior to matter from the fact that matter accepts form and is not opposed to it, while privation, being opposed to it, does not persist with it.[366] It is clear that the unlimited in matter is privation from the fact that form is the limit of the compound; that is why nature is stable when it meets with it. But privation is contrary to form and as limitlessness is to limit. That is why it does not persist with form, so that the being of limitlessness is in privation, when the unlimited is viewed not as being inexhaustible through being privation and matter, but as being formless and unshaped and thus having no limits, and, as being potential, thus also limitless – features that are present through lack of form. However, when it has received form, its limitlessness and incompleteness comes to a halt. Thus I think it possible to say that privation is not simply limitlessness, since the concept of privation is one, that of limitlessness another; but that absence of limit is present in privation as limit is in form. But perhaps someone will say, speaking well, I believe, both that limit is a completion of form and is not added to it as coming from somewhere outside, and that limitlessness, which is of a kind to be called privative and not form, is a completion of privation. For this absence of limit is a privation of a formal limit, and privation is such as the absence of limit and form in what is of a nature to receive them. Except that as privation is contrary to form, so is the unlimited to limit.

But what about what comes next which says: 'the substrate as such is the continuous and perceptible'? For if matter is the substrate of privation and limitlessness, but the substrate as such of privation, and the limitlessness of division without limit is the continuous and perceptible, i.e. the natural body, it seems evidently to follow that matter is body, which some call the second substrate. Perhaps 'as such' is not joined to 'substrate' idly. For it is not matter alone that is substrate to division without limit and privation as such, but rather that is already continuous perceptible body. For what is to be divided must be the actually extended and not merely the potentially so, like matter. So what does it mean to say of matter that it is the unlimited in division? Perhaps that when the compound body is cut without limit, since the cut is not in form, (for the thing cut remains water or air or flesh or bone) it is necessary that the cut be with respect to matter. So as primary limitlessness, which is spoken of with regard to the limit of the simple form, consists in primary privation with regard to primary matter, so limitlessness in division is in the privation in the perceptible body. For that is what is divided.

But how is privation the being of the limitless, as he says, but as a cause the limitless is called matter and not privation? Perhaps it is because privation was viewed as involved with matter among causes, since it is not a cause as such but incidentally; for it is so through absence. So reasonably privative cause is counted as material. But if continuity is form, while division and cutting ends continuity, how can we say that form remains in division? Perhaps it does not end continuity, since the segments remain continuous as well, but only a certain quantity of continuity. But then it also ends the form as so much; so if the division is in this respect it will be with respect to form and not to matter. But perhaps it is not with respect to a quantity as a formal element, for as formal a quantity is also indivisible, but only with respect to material separation which has turned aside from the indivisibility of immaterial simplicity.

Having shown that being unlimited belongs rather to the material cause as well as does privation, he again confirms the argument also from the opinion of the others who suppose the unlimited to be existent and a principle. For those people supposed that the unlimited was matter, some saying that it was air or water, others an intermediate; and they regarded the coming to be of other things from this not as from an efficient but as from a material cause. For they say that things that come to be do so from this as primary, and they say that they are resolved into this at the end. But he accuses them of contradicting themselves, since they hypothesize the limitless as containing. For those who say that limitlessness is in matter ought to say that it is contained rather than that it contains.[367] For matter is contained and determined by form. Though perhaps they said that it contains as being seen in all bodies and characterizing all material things.

Chapter 8 begins here in modern editions.

208a5 It remains to rehearse the arguments through which the unlimited seems to exist [not only potentially but as a separate entity. For some of them are not conclusive and others meet with other true responses. For it is not necessary, in order that generation should not give out, that there should be an actually unlimited perceptible body; for the end of one may be the birth of another,] while the universe is limited.

He has shown that the unlimited is potential and not actual, and has articulated all the problems belonging to this discussion. Next he wishes to dissolve all the arguments that have been put forward which supported an actual unlimited. There were initially, if we recall them, five arguments that posited the unlimited; 1. from time, 2. from the division and addition in magnitudes, 3. because only in this way would coming and ceasing to be not be exhausted, if there were an unlimited source from which to take what comes to be, 4. because the limited always is limited by something, 5. and especially and most importantly, because the conceived is inexhaustible. Of these five he himself showed that the first two necessarily posited the unlimited; for these posit the potential and that which has its being in becoming, the unlimited and that which proceeds without limit. The rest he will try to refute, being the ones that posit it as separate and actual. For thus the argument would be complete, if we did not only prove what was to be proved but also refuted those who held the opposite view.

So he says that of these arguments one will be refuted, as uselessly saying that the unlimited was necessary for inexhaustible generation; others will be met by more true arguments that reject them. For what cannot be otherwise is necessary; but generation can be inexhaustible without there being an actual unlimited. For it is possible for the end of one thing to be the beginning of another, the whole being limited in magnitude, and if that were so coming to be would be inexhaustible. For what would bring such a system to an end? And we shall have no need of the actually unlimited. So the argument is not compelling, since it takes as necessarily following what does not invariably follow. For even if coming to be is inexhaustible there need not be an unlimited magnitude in actuality. So such a conditional is neither sound nor compelling. So also the additional assumption[368] and the supplement added to it, which says 'but generation is inexhaustible and therefore there is an actual unlimited', do not later follow necessarily.

208a11 Also to touch and to be limited [are different; for the former is relational and has an object. For everything touches something and this is incidental to something limited. But to be limited is not a relation.] Nor can anything whatever touch anything whatever.

This is the fourth of the arguments that were set out there, which says that there is something beyond everything that is limited, at which it is bounded, and introduces the actual unlimited because this always comes about. For if there were no actual unlimited there would not always be something beyond the limited at which it was bounded, but we should meet with something limited which was not bounded by something else. But it is possible truly to object to this argument by saying that being bounded is different from touching. For to touch is a relation, since what touches touches something touched which is other than itself; but being bounded is in the category of quantity.[369] So, if each belongs to a different kind, being bounded does not consist in touching something. For being bounded is not a relation but a characteristic. Also some limited things can both be bounded by something and touch something. But that is an attribute of such things as are not simply wholes but also parts of the universe; but this is not through having a limit but because they are parts together with other parts. But if some single thing is a whole and that is everything, why is it bounded by something else? But those who say these things appear to have been misled by things that touch and think that touching and being bounded are the same thing. But if someone were to say that a limit is relative, since a limit is a limit of something, still it is not related to something else, but to itself. For a limit is of what is limited, and not of that thing at which it is bounded. Then he adds another argument to the account, one distinguishing being bounded from touching. For if it is possible to be bounded by everything, but not to touch everything, it is clear that being bounded and touching are not the same thing. For a number such as ten and a sound separated off by silence are bounded, but what would they be said to touch? Surely not silence! But also a limited line on a surface is bounded by points, but is not said to touch anything, unless one were to add another line to it. For not anything is of a nature to touch something. For a sound does not touch a line, although each is bounded.

It seems surprising that Alexander should say: 'It is not true that the limited has a limit *qua* being bounded. For a number at least is bounded, but it does not have a limit.' But how can something be bounded if it has no limit? Rather, he appears to seek for a limit of the limited based on similarity to magnitudes, one which is not a part of it but is at an immediate remove. But nothing prevents the units at each end being limits of the number, in one respect being parts, in another respect limits, by which they are analogous to points. At any rate a point is said to be a unit that has a position.[370] But perhaps a number is itself as such a limit of a definite quantity, and one should not look for another limit of it.

208a14 But it is absurd to trust in thought. [For the excess and deficit are not in the object but in the thought. For one could think of each of us growing many times bigger without limit. But nothing extends beyond the city for that reason, or is of such a size as we have for the reason that somebody thinks of it, but because it is so.] But that is a coincidence.[371]

The fifth argument was that from thought and imagination, which, he says, especially leads to positing an actual unlimited. For because they are never exhausted in thought, number seems to be such, and also mathematical magnitudes, and the area beyond the heavens.[372] He refutes this argument concisely and clearly. Such excess as this, he says, is not in objects, but the objects are objects of thought. For if we think and imagine and dream of something it does not also thereby exist. For in that way the goat-stag and many other things would exist, of which none either does or could exist. For as artists think of and paint many things, not only species but also small instead of big and big instead of small, so imagination paints and contrives in itself tens of thousands of things that do not exist. Thus it is possible to imagine Diares as a thousand miles tall, or bigger than the city, with what of him that is bigger being outside the city.[373] For I think it is better to understand 'beyond the city' in that way, which is how Eudemus understood the words of his master, as being of people imagining Diares as bigger than the city, and not as Alexander explains it, as when we who are in the city imagine ourselves as being outside or many times bigger in magnitude, so that 'beyond the city' illustrates only a place. Also he appears to explain it thus through having understood 'but nothing is outside the city for that reason' and 'of such a magnitude as we have' as referring to two different things; each is said about increase of magnitude, either a magnitude exceeding the city or simply the size that we have.[374] For in that way also the fantasy will be relevant to that about the unlimited. For the imagination of something as being in a different place only shows that imagination alters things, but does not show it as being relevant to the unlimited.

So things are not in such a way because they are thought of as being in that way, but because they are so. Their being thought of, whether as they are or as they are not, happens to them incidentally. So even if things are thought of as they are, they are not so because of being thought of, but rather they happened incidentally to be thought of. So if being thought of is also incidental and thought does not always chime with the facts, one should not make statements about facts on the basis of thought or imagination. Thus nothing is outside the heavens because it is thought of, since they are the totality and the whole.

208a20 But time and change are unlimited [and thought also, without any given part being permanent. But magnitude is unlimited neither through exhaustion nor through increase in thought. But the discussion about the unlimited is finished, in what way it exists, in what way it does not, and what it is.].

518,1

Having refuted those of the arguments positing the unlimited which introduce an actual unlimited, he next <discusses> the first two, that from the limitlessness of time and that from the unlimited divisibility of magnitudes.

(4)

LACUNA

... he evidences <from> their limitlessness and from the thought of them. For the thought of it also is not as of a simultaneous whole nor as of any given part as being permanent, but it is given as what has its being in becoming. For in general, if one embraces it as a whole, one sets a limit about it. But addition is to what is always limited. But it is absurd to say that also in thoughts of the eternal <and divine> ... are not permanent ...

15

LACUNA

... for it would both be composed of unlimited parts and actually unlimited. However, the division into the sections already taken is not compelled to come to a stop. But as to the way in which it exists and the way it does not, <he has shown> that it is not actual and a simultaneous whole, but it is potential part by part. And, as to what it is, that it is that of which, quantitatively, there is always possible to find something beyond ...

(26)

30

Notes

Notes written by Urmson are marked (JOU). In a few cases notes are co-authored.

1. 192b21-3.
2. Aristotle defines natural things as having the principle of motion and rest in themselves in *Phys.* 2.1, 192b13-5.
3. By referring to bodily change Simplicius makes a restriction as he elsewhere admits the existence of the change (or motion) of the soul, see 1249,1-4. He explicitly follows Plato (*Laws* 10, 894B8-C8; *Phaedrus* 245C5-246A2) in that.
4. The reference is to the Stoics, see *SVF* 509, 510. Later on, in 700,21-8, Simplicius ascribes this view explicitly to the Stoics (see also *in Cat.* 350,15-16)
5. This is the account of the *Timaeus* (37D5-7), taken over by Plotinus (e.g. 3.7.1, 19-20) as well.
6. The passage has not been included in Diels-Kranz, but see A 37 (which is Simplicius *in Cael.* 294,33 ff.) and B 9.
7. Discussion of continuity is to be found in *Phys.* 6.1-2.
8. The lemma gives only the first few words of the passage under discussion in this case; usually the closing words are also given. The text in square brackets is added for the convenience of the reader. (JOU)
9. This is the reading of F and the Aldine edition; Diels' text has 'change' and 'transformation' in the reverse order, apparently in agreement with most MSS. But Aristotle says often in Book 5 that transformation is wider than change, since, in his terminology, change is always in a persisting substance, but coming to be and ceasing to be are not changes in a persisting substance. See e.g. 225a25-6 – coming to be is not change – and 225b23 – change is in quality, quantity or place. But changes are also transformations, as well as coming to be and ceasing to be. The word usually translated as 'change' is *kinêsis*, which is always so translated here in statements intended to hold of all change, even when the illustration is one of local motion. It is translated as 'motion' only when it is used exclusively of that species of change. The word translated as 'transformation', somewhat artificially, is *metabolê*. *Kinêsis* in non-philosophical Greek normally means motion, except for transferred uses to refer to civil disturbances and the like. *Metabolê* is usually translated as 'change'. But Plato at *Theaetetus* 181C-D lists change of colour, growing old, and other alterations as being *kinêseis*; in *Laws* 894A he lists coming to be as a *metabolê* which is included in a list of ten kinds of *kinêsis*, thus treating *metabolê* as a species of *kinêsis*. Aristotle in *Categories* 15a13 also lists six kinds of *kinêsis*, including coming to be and ceasing to be, in agreement with Plato. But in Book 5 of the *Physics* Aristotle insists that *metabolê* is a genus which includes coming to be and ceasing to be as one species which is not a *kinêsis* and several species which are. It seems clear from this that philosophical use of these terms is varied and somewhat arbitrary, which must excuse the arbitrary use here of

'transformation' and 'change'. In this Book, unlike Book 5, they mark no difference. (JOU)

The two translations – 'transformation' and 'change' – have been used consistently with the translation of the commentary on Book 5 in this series. Simplicius is well aware of Aristotle's usage in Book 5. He writes at 801,1: 'In the third book of this work he has taught about change in its widest sense, which is the same as transformation, and has used the words 'change' and 'transformation' for it without distinction. Now he wishes to distinguish them and to show that transformation is more general and more universal, while change is more particular.' If he wrote the text here as in Diels, it can only have been a slip of the pen. What is clear is that both in the Greek and this translation the use of the terms is artificial. (PL)

10. 225a7-b10.

11. In fact, Aristotle discusses change as one of the postpredicaments (*Cat.* 14, 15a13 ff.), not as something having to do with relation. Neither does Simplicius subsume it under relation in the commentary on the *Categories*. Such an interpretation might go back to the Peripatetic Boethus (ap. Simplicius *in Cat.* 433,32-434,17). But Simplicius contests that view (*in Cat.* 434,18-19).

12. This is a general principle of Aristotelian methodology, cf. *Phys.* 1.1, 184b16-21; *DA* 2.2, 413a11-12; *NE* 1.2, 1095b2-4.

13. Features *kath' hauta* are thus twofold: (a) either the essential features which complete the determination of the thing (see also 123,13-14), or (b) the essential attributes which also mark off the species from another species in the same genus, but are not included in the definition. In humans, 'rational' exemplifies (a), 'laughing' goes with (b). But the example he gives – on fire being active – does not fit into the frame, for 'active' is not part of the definition of fire nor a distinguishing mark of it.

14. 4b20 ff.; Simplicius repeats it in 81,17-18.

15. The passage has not been noted in D-K. For the notion that void is necessary for there to be motion, see Melissus B 7, Democritus A 58 (ap. Simplicius *in Phys.* 1318-24), A 165 (ap. Alexander of Aphrodisias *Quaest.* 2.23, p. 72, 28-9 Bruns). The definition that void is a place deprived of body reflects Aristotle's ideas in *Phys.* 4.1, 208b26-7, see also *Cael.* 1.9, 279a13-14 and *GC* 1.8, 326b19.

16. According to Book 5 coming to be is not a change but a different species of transformation. (JOU)

17. This is the definition of time given in Book 4. Roughly Aristotle means that time is that element in change which is counted. (JOU)

18. Black and white are the basic colours from which each colour is to be derived, see *Sens.* 3.

19. i.e. the heavenly bodies called etherial as being in the ether, the fifth element. (JOU)

20. The argument seems to be this. Colour is a quality and being visible is a relation requiring a living being who sees and an object to be seen. But colour can also be taken to be substrate/subject (*hupokeimenon*) of the relative term 'being visible'. Similarly, snow is also substrate/subject for the term 'being meltable' – though snow is not a quality. But if we equate snow with being meltable then we also have to equate being divisible and being meltable for, as a physical object, snow is divisible. But this is absurd. If so, we have to reach the conclusion that there is nothing which is only potential in existence.

21. The argument is difficult to understand. It seems to suggest that, though snow is meltable and the meltable is divisible, still all these must be distinguished unless we are to reach the conclusion that being divisible and being snow are the same thing.

Notes to pages 16-19

22. Reading *mêden* at 399,3 instead of *mêdenos*.
23. *Enn.* 2.5.1, 17-21 – up to 'in general' – and then 11-16. Simplicius thus inverts the order of Plotinus' argument and quotes a version which is fairly different from the text we have in both editions by Henry and Schwyzer. E.g., the last sentence is missing from the texts by Henry and Schwyzer.
24. *Enn.* 2.5.5, 34-5.
25. The way Simplicius is interpreting Plotinus is a bit idiosyncratic. The primary existent can be regarded as form only if we assume that the One is beyond all kinds of matter, physical and intelligible alike. But the potential is present in the intelligible world as well (e.g. 2.5.3, 3-4). On the other hand, as potentially everything (*dunamei panta*, 2.5.5, 5) matter is non-existent in a way.
26. Alexander's punctuation has been followed by Bekker and W.D. Ross as well.
27. 201a9.
28. Porphyry, *Test.* 151 Smith. Thus, on interpreting the passage (200b26) Porphyry dissents from Alexander (and modern commentators such as Ross and Hussey) insofar as he thinks that due to the lack of all kinds of determination the potential cannot admit division along the lines of the categories. To make this point he may also have to excise 'and in actuality' after 'potentially'. In any case, as it stands the testimony seems self-contradictory because of two incompatible claims. The first is that Aristotle did not divide up the potential but identified it with prime matter, while according to the second, Aristotle may have said that the potential is tenfold. One way of getting rid of the confusion is to assume that the quotation ends up at 'tenfold' at 399,27. Hence, 'Also perhaps' signals the beginning of Simplicius' comment.
29. For Themistius, see 67,20-2 Schenkl (*CAG* 5.2).
30. That is, unlike Ross' text, this version keeps *ti* ('something') and puts a comma before and not after *kai entelekheiai* ('and in actuality').
31. 201a9-10.
32. Place (*to pou*) belongs to physical bodies, time (*to pote*) is the number of motion (219b1-3), posture (*to keisthai*) may be akin to the relatives (*Cat.* 11b10-11). Simplicius takes the terms from Aristotle's *Categories* and his remark is reinforced by Plotinus' conviction (*Enn.* 6.1-3) that Aristotle's categories apply to the physical world only.
33. *Timaeus* 57E.
34. The point here is that as in English words ending in 'er', such as 'doer' and 'striker' are frequently indicative of being active and not passive, as is 'sufferer', so, in Greek, words ending -*ikos* are normally active, but the word *pathêtikos*, here translated 'sufferer', is passive in meaning. (JOU)
35. This also might be a reference to a Platonic doctrine, at least Platonist sources attribute to Plato the view that the 'undefined duality' (*aoristos duas*), the material principle of the world, is sometimes called 'great and small', see, e.g. Aristotle, *Metaph.* 1.6, 987b20-1; Alexander, *in Metaph.* 56,2; 117,26-7; Philoponus, *in Phys.* 521,12.
36. *Sophist* 248E7-249B10 (change is one of the highest genera of being); *Parmenides* 138B6-139B3, 162E1-163B6. But these passages do not demonstrate that Plato differed from Aristotle at this point, see W.D. Ross in his commentary on Aristotle's *Physics* (Oxford 1936), 536.
37. *Phaedrus* 245C5-246A2.
38. The text reads *epei mê estin exô tôn pragmatôn*. My translation requires something like *ei estin* ..., and this may not be plausible, but it makes better sense than 'since it is not outside things'. (JOU)

39. Plotinus applies the concept of motion and rest to the activity and structure of the Intellect, see, e.g. *Enn.* 5.1.4, 35-7.

40. Diels has a comma before 'and place' and a full stop after 'of place'. But this obscures the point that changes in quality and in quantity, for example, are changes of things different in kind, whereas changes of quality in different places will be 'even in the same things'. (JOU)

Simplicius refers to the same explanation by Alexander at his *in Cat.* 31,23 ff. Accordingly, homonyms are of two types, one is incidental (both Paris and the king of Macedon are called Alexander), the other is intentional. (PL)

41. 1a6.

42. An *arkhê* may be a beginning in time, a logical starting point or principle (as often in this work), the government, and no doubt other things that are in some way primary, but not in the same way. They are presumably named analogically, like the foot of a man and the foot of a mountain.

43. The sentence reflects Simplicius' conviction that the doctrines Aristotle and Plato are reconcilable 'in most cases', see *in Cat.* 7,23-32. The conviction originates in Porphyry, see his *in Cat.* 57,7-8; 58,5-7; 91,19-27, and P. Hadot, 'The harmony of Plotinus and Aristotle according to Porphyry', in R. Sorabji (ed.), *Aristotle Transformed. The Ancient Commentators and Their Influence.* London-Ithaca, NY, 1990, 125-40. The argument on the five highest genera of being in Proclus may refer to an exegesis on Plato's *Sophist*. Proclus wrote a commentary on the work but it has been lost. On Proclus' views on that dialogue, derived from extant writings, see C. Steel, 'Le *Sophiste* comme texte théologique dans l' interprétation de Proclus', in E.P. Bos & P.A. Meijer (eds), *On Proclus and His Influence in Medieval Philosophy*. Leiden 1992, 51-64.

44. For the notion in Proclus, see *Inst.* § 176 and *in Tim.* 2, 254,6-28.

45. Strictly speaking, only the ideas can exist on their own, while particulars are related to other particulars in a disordered way, just as they are related to themselves in a disordered way, see *Phaedo* 73E3. But the assumption is quite common in Plato.

46. Simplicius is referring to the *pros-hen* relation ('focal meaning') and it has been argued that the relation was worked out in the Old Academy, by Xenocrates, see H.J. Krämer, 'Zur geschichtlichen Stellung der Aristotelischen Metaphysik', *Kant-Studien* 58 (1967), 313-54.

47. 248E7-249B6. Divergences from the new OCT text are considerable. When the argument talks about vital and intellective (*noeros*) powers (potentialities, [*dunameis*]) and activities, Simplicius rules out that it can be about the intelligible (*noêtos*) realm as well. Intellective entities are the divine beings and the different souls, while intelligible being belongs to the Intellect-hypostasis where there is no place for potentiality.

48. In 406,2, the text reads *kinêta* 'changed' which is absurd. The received text of Plato has *akinêta* 'unchanged' which I have translated. I have added the speakers' names. (JOU)

49. Where *ekei* 'there' refers to the eternal, intelligible world it is capitalized; similarly where *enthade* 'here' refers to the world of change it is capitalized as HERE. (JOU)

50. Plato did not say this. What Simplicius gives here is a blend of the theories in the *Sophist* and *Timaeus* interpreted by Proclus to whose conception Simplicus is referring again. In *Proclus*, see *Theol. Plat.* 3.19, 64,20-65,13 S-W.

51. Paradigmatic causes are the ideas which serve as paradigms for the Demiurge to arrange the physical world, see *Timaeus* 28C6. By the term 'project'

Simplicius alludes to the theory according to which human soul projects *logoi* (concepts which can also be active forces) which are rooted in the Intellect.

52. Porphyry, *Test.* 152F. Smith.

53. I fail to understand this. Aristotle does contrast the incomplete with the complete as being lesser in the lemma. But there is no mention of the moderate in the lemma. Nor could this moderate be an intermediate, as grey is between the contraries black and white in Aristotle's theory of colour, and is said to be opposed to both (see 406,26 ff.); for the complete and incomplete are contradictories and thus have nothing intermediate. Nor could the moderate be less than the incomplete, for even a second is an example of an incomplete century. I have no idea what the moderate could be. (JOU)

One might take 'opposite' in a weak sense, as the ethical mean can be said to be opposed to both extremes; but these extremes are, of course, contraries, not contradictories. One might say that Aristotle applied complete and incomplete to quantities, as the lemma says: 'and in quantity, since one is complete the other is incomplete'. And quantities allow intermediaries and this kind of complete and incomplete allows degrees. (PL)

54. There is no example in *LSJ* of the verb *analogein* being used transitively as it appears to be here. (JOU)

55. In Book 5 he says that there is transformation, not change, in the category of substance, because coming to be and ceasing to be are not changes, which occur only in persistent substances.

56. The reference is not to the *Categories*, rather to *De Int.* 9, 19a9 ff. See also *Phys.* 5.1, 225b5.

57. These are examples of alleged change in the categories of relation and of posture. (JOU)

58. *Phys.* 5.1, 225b2 ff. Aristotle wavers as to whether substantial transformation (*metabolê*) is not to be separated from the other three changes (*kinêseis*). Simplicius reflects on the problem at his in *Phys.* 823,5 ff.

59. Simplicius is quoting *Phys.* 5.1, 225b10-13.

60. These quotations from 225b10 ff. are not verbally exact or complete, though near enough. (JOU)

61. *Cat.* 7, 6b15-17. But Aristotle does not accept that all kinds of relative have an opposite term, cf. 6b17-19.

62. 3.2, 202a7-8.

63. *Cat.* 9, 11b2-4.

64. Fr. 59 Wehrli. As Aristotle defines time as a property or number of change, the notion of change in time would imply the absurdity of change in change. This is what Simplicius and Eudemus see.

65. This, together with being shod, is one of Aristotle's first examples of the category of state (*hexis*) in the *Categories*. (JOU)

66. *here* means the physical world as opposed to the intelligible realm.

67. Grammatically Simplicius says that the substrate comes to be and ceases to be in these respects while the substrate continues. (JOU)

68. Fr. 59 Wehrli. See also 411,17.

69. *Test.* 153B FHSG.

70. *Test.* 153B FHSG.

71. Simplicius omits here the verb *einai* which we have in the lemma. This is why the translation of the quotation slightly differs from that of the lemma.

72. 'Black' and 'white are used here for consistency. 'Dark' and 'light' would be more natural.

73. We do not have the commentaries by Alexander and Porphyry. In Smith's edition of Porphyry's fragments and testimonies this is *Test.* 153F. For Themistius, see his *in Phys.* 69.6-7 Schenkl.

74. Porphyry, Test. 153F.

75. 201b31-2.

76. As opposed to coming to be of a certain sort. (JOU)

77. 'ripening and aging' (*hadrunsis gêransis*) is the order of the received text of the *Phys.* 201a19, but here as at *Metaph.* 1065b20 (noted by Diels). (JOU)

78. By 'back-reference' Simplicius means a kind of paronymy (see 210,16; 363,20) whereby we can transfer names from the genuine bearer of the name to things/properties which are connected to the genuine bearer.

79. Aristotle does not use the word 'suitability' (*epitêdeiotês*) but Simplicius feels free to apply it because in this way he can link Aristotle's remarks to Neoplatonic theories according to which things are suitable to receive forms or *logoi* sent by the Intellect or the World-Soul. In order for matter to be suitable to receive the proper form it has to receive a primary influence from the World-Soul, which is called *emphasis* (e.g. 94,7-8).

80. e.g. 5.1, 224b7-10.

81. As Diels prints the text, there is a very abrupt anacoluthon at 417,15. There seems to be a lacuna, perhaps to be filled by some such phrase as 'is mistaken'. (JOU)

82. Accenting *tinos* and *ti* which Diels leaves unaccented at 417,16. (JOU)

83. In that case, building will be coming-to-be of a house, not an increase of the number of bricks, while growing old will be an increase or diminution of the proper ingredients of the body.

84. Darii, in the first figure of the syllogism, of the form: If all M is P and some S is M then some S is P. (JOU)

85. Plotinus downgrades matter at 6.3.8 (19-37) but insists it exists at 1.8.15, 1-2. Proclus says (*in Tim.* I 226.20-1) that Neoplatonists are accustomed to call matter non-being.

86. At 419,1-2 Diels prints this as a quotation, but the actual words do not occur in Aristotle. Cf. 201a27ff. (JOU)

87. The reference may be to *DA* 3.5, 430a17-18 where Aristotle discusses the active intellect which is separable, impassible, unmixed as essentially actual. By 'intellective' (*noeros*) Simplicius refers primarily to the soul and other spiritual beings. The term does not hint at the entities in the intelligible realm; they are called intelligible (*noêtos*).

88. Of the three dispositions of the heavenly bodies, the first, which issues from their co-presence, refers to the presence together of heavenly bodies (see Petosiris ap. Vettius Valens, *Anthologium libri* 80,4 Kroll and Philoponus, *in Meteora* 48,29), the second may refer to the 'rectangular position' (*tetragônikê stasis / diastasis*) of the sun and the moon (see Simplicius, *in De Caelo*, 480,4-5), to the division of the zodiac into four equilateral triangles (see Ptolemy, *Tetrabiblos* 1.18), and the hexagonal (*hexagônikê*, not noted in the most recent edition of *LSJ* [1996 with supplement]) refers to the sextile aspect (the aspect of two heavenly bodies which are 60° or one sixth part of the zodiac distant from each other, see Vettius Valens 102,8), while the third comes from the 'conjunction' where a heavenly body is not in aspect with other heavenly bodies, see Vettius Valens 102,8. (for the term itself, see *Phys.* 1.7, 191a2 and *Phaedrus* 247C6).

89. The reason may be that all are made of the same element, the aether.

90. The reference is to Book 8 where Aristotle is about to demonstrate the existence of an unmoved mover.

91. For a Stoic definition of body as what can act and be acted upon, see Jaap Mansfeld, 'Zeno of Citium', *Mnemosyne* 31 (1978), 134-78 at 160-1.
92. This is the view of, e.g., Empedocles, Diogenes of Apollonia and the atomists.
93. 139A2-3.
94. Reading *sunkhusin* with Giacomini at 420,19 instead of *sunkrisin* – 'composition'. (JOU)
95. As ever, Simplicius is quoting from Alexander's commentary on the *Physics*, now lost. But we should bear in mind that in his *Quaestiones* 1.1, 4,1-4 he argues that the first mover moves the divine sphere by being the object of its thought and desire. The other way to explain the difference is to assume that this text of the *Quaestiones* is by another Peripatetic.
96. 28A2, 29A1.
97. 897A8-B2. Simplicius omits 'in the same spot' (*en tôi autôi*) in A8.
98. This is also attested in *Phaedrus* 245C5-8.
99. *Cael.* 2.12, 292a18-21. Simplicius omits *monon* in a19: 'we think about them as *nothing but bodies*'.
100. *Phaedrus* 245C5-8.
101. Simplicius omits 'earth and sea' after 'heavens' and reverses the order of 'forethought' and 'consideration'. (JOU)
   Cf. *Laws* 10, 896E8-897A2. Plato here calls the soul's changes *kinêseis*, but Simplicius is right because Plato distinguishes primary (*prôtourgoi*) changes from the physical changes that are secondary (*deuterourgoi*) in 897A4-5. (PL)
102. The soul moves along with the body which it belongs to, see *DA* 2.1, 415b15-17.
103. Such an order of changes does not come from Plato directly. Rather, it reflects Middle- and Neoplatonic systematization. As for the lowest level, Plutarch of Chaeroneia (*De animae procreatione in Timaeo* 1015E) and Atticus (frr. 23 and 31 Des Places) say that before the divine intellect made this ordered universe, the precosmic stuff was in disordered motion (see also Galen, *On the formation of the embryo* 4, 696 Kühn). The fourth stage reflects the Neoplatonic view that the hypostasis Intellect has its own motion. Relying on the Sophist they apply the five highest genera of being to the structure of the intelligible realm (e.g. Plotinus 5.1).
104. Ross reads *tou dunamei ontos <entelekheia>* – the actualization of the potentially existent; Simplicius apparently understands *energeia* as the subject. It makes no difference to the sense of the passage. (JOU)
105. See Themistius, *in Phys.* 70,33-71,1 Schenkl. The text is remarkably different from that which Simplicius attributes to Themistius. Simplicius either distorts Themistius' version deliberately or he had access to another copy of Themistius' paraphrase. The latter version may be less plausible for the differences are huge. The commentary by Aspasius, a Peripatetic philosopher of the second century AD, has not survived.
106. Porphyry, *Test.* 154F Smith.
107. Aristotle holds that *phora* – here translated as 'travel' – can be used only of the inanimate; natural travel is that upwards or downwards to its proper place; the movement of animals is internally caused for quite different reasons. (JOU)
108. For the Neoplatonists, nature as such has no independent existence. It originated in the soul hypostasis and therefore is incapable of changing by itself – change requires soul.
109. According to the text, the lemma ends at 'one and the same' and does not contain the last sentence. But the last sentence is included in this piece of commentary and the next lemma begins after it. (JOU)

110. The syllogism is in Camestres, All P is M, No S is M, so No S is P. The order of the premisses is reversed. (JOU)

111. 201a19-21.

112. This fits the Aristotelian principle that in pursuing studies we have to start from what is obvious to us, see *Phys.* 1.1, 184a10-21.

113. By 'juices' Simplicius may mean the four Hippocratic humours (blood, phlegm, yellow bile and black bile), on breath, see the treatise *On Breaths* which also emphasizes the primacy of blood (14,4-19). Alcmaeon thought of disease as imbalance between opposite powers (Aëtius 5.30,1 = 24 B4 D-K). The importance of the most immediate objects in curing disease was stressed in *On Ancient Medicine*. It is interesting to know that, as far as our sources allow us to say, at least some of the representatives of Hippocratic medicine criticized the view that disease is to be brought back to primary elements and qualities, such as wet, dry, cold and hot, see *On the Nature of Man*, 1,1-19 and *On Ancient Medicine* 15,5-13.

114. *DA* 2.7, 418a26-b2; *Sens.* 3, 239b11-12. The first account tells us that colour along with glow is one of two things visible in the strict sense (*kuriôs*) explained in 2.6.4, 18-24. The nature of colour is to produce a certain kind of change in the illuminated transparent medium. Glow, by contrast, can be seen in the dark. The second definition exploits the idea that even opaque solid bodies contain some degree of transparency, since they are made partly of water and air, and their colour is not the surface of the body, but the surface of the transparency in the body. (Ed.)

115. Attribute of colour (*sumbebêkos*). Simplicius means that the first account of colour (in the previous footnote) does not make it a matter of definition that colour is visible, but a non-definitional attribute, even if a necessary one. In my view, Aristotle does not want to make visibility definitional of colour, because in 2.6 he has taken the opposite route of making colour definitional (*kath' hauto*, 418a8-9; 24-5) of vision. 'Simplicius', *in DA* 130,11-12 agrees that visibility is not definitional of colour (Ed.).

116. 'Not a relation but self-existent': Simplicius' point at 425,31-3 is that Aristotle's definition of colour in *DA* 2.7 is not that it is (like glow) visible, but that (unlike glow) it changes the illuminated medium. Therefore Aristotle does not define colour as something relative to vision. (Ed.)

117. Diogenes of Babylon, a disciple of the Stoic Chrysippus, wrote a treatise *On Music* that influenced the Epicurean Philodemus when he wrote his own *On Music*. Diogenes' treatise is lost, but fragments of Philodemus' work have survived and serve as a basis for reconstructing Diogenes' text (see *SVF* III.II.54-90).

118. Alexander gives a slightly different definition of voice in his *in De Sensu* 66,15-17: e.g., instead of 'deliberate beating' (*proairetikê plêgê*) he speaks of 'beating accompanied with *phantasia*'.

119. Added to 'potential' at the end of the lemma at 201a29-b5 above. (JOU)

120. This reflects Simplicius' theory that to receive form, the body must first receive suitability (*epitêdeiotês*) to receive that form. Cf. note 79 above.

121. Omitting *dunamei* at 427,17. (JOU)

122. Themistius, *in Phys.* 72,2-7 Schenkl, Porphyry, *Test.* 155F Smith.

123. Plato never defined change in terms of departure from being (*exanastasis apo tês ousias*). In 405,23 ff. Simplicius believes he can prove a similar thesis ('first departure of being', *prôtê exanastasis apo tou ontos*) by drawing on *Sophist* 248E.

124. The main point is that Aristotle distinguishes activities (*energeiai*) that are complete in themselves, like thinking, from *kinêseis*, like building a house, which are not complete until finished. He therefore calls *kinêseis* 'incomplete activities'. For Plato, by contrast, *kinêsis* is one of the five Great Kinds in the *Sophist*, and activity (*energeia*) is merely one of its species.

125. For the columns, see Pythagoras 58B5 (=Aristotle, *Metaph*. 1.5, 986a22-6). There are other columns as well, some containing less than ten pairs of opposites, e.g., in Philoponus *in Phys*. 124,5-12.

126. To refer to the status Simplicius uses the term *parüphistanai*. The term, just as the noun *parüpostasis*, refers to a kind of existence which is parasitic. This kind of existent is not only dependent on another, *per se*, existent but also derived from it as parasitic. In this form, the theory was introduced by Proclus, see A.C. Lloyd, 'Parüpostasis in Proclus', G. Boss & G. Seel (eds), *Proclus et son influence. Actes du Colloque de Neuchâtel*, juin 1985. Zürich 1987, 145-57, and J. Opsomer & C. Steel, 'Evil without a cause. Proclus' doctrine on the origin of evil, and its antecedents in Hellenistic philosophy', Th. Fuhrer & M. Erler (eds), *Zur Rezeption der hellenistischen Philosophie in der Spätantike*. Stuttgart 1999, 229-69, esp. 249-52. For the notion of *parüpostasis* in this commentary, see 250,21; 1262,8.

127. *Sophist* 256D11.
128. *Sophist* 250A4 ff.
129. *Sophist* 256D11-12.
130. *Sophist* 256D8-9.

131. For the Pythagoreans, see 201b20 and *Metaph*. 11.9, 1066a10. However, Aristotle says only that change is according to inequality and the non-existent while Simplicius takes him to equate change and the non-existent.

132. Fr. 60 Wehrli. Cf. M. Isnardi Parente, *Atti della Accademia Nazionale dei Lincei* 395 (1998).

133. Archytas 47 A 13 D-K.

134. Diels marks *alla gar hôristai ouk esti* at 431,15 as defective. I translate Diels' conjecture *alla gar horistê ouk esti*. (JOU)

135. Supplying *loipon* at the beginning of 432,10 with Diels. (JOU)

136. For Plotinus, change is one of the highest genera of being. In this he followed Plato's *Sophist* and applied these genera to the intelligible world while using Aristotelian categories to describe the physical world.

137. *Timaeus* 57E2-58A4.
138. *Timaeus* 58E2-4.

139. The quotation is in 431,8-12. The reference to Eudemus is fr. 60W (though the repetition at 433,13-16 is not included in Wehrli's collection).

140. This reflects the principle that the efficient cause is greater than its effect, see, e.g., 314,9-23.

141. Aristotle does not allude to it. Simplicius refers to the column (429,9-18) where change (or, rather, things changed) is on the privative side.

142. 201a23-4.

143. In his extant works, Alexander does not use the term 'intellective substance'. In the Neoplatonic context, it refers to divine beings and separate souls. It is distinguished from 'intelligible' which refers to the contents of the realm of the Intellect. The distinction is very well attested in Proclus.

144. Simplicius traces back this view to Plato in 821,21-7.

145. The quotation from Aspasius reads, as punctuated by Diels, as follows: *kineitai de kai to kinoun, hôsper eirêtai, pan to dunamei on kinêtikon*, I translate it as punctuated *kineitai de kai to kinoun hôsper eirêtai pan, to dunamei on kinêtikon*. I agree with Ross that *hôsper eirêtai* specifies the natural as opposed to the unchanging changer in Aristotle's text, and I so take it here. Simplicius at 434,33 and 436,22-4 appears to agree with this reading. The difference of Aspasius' reading from the accepted one is that it says 'being the potential initiator of

154   *Notes to pages 54-61*

change' instead of 'being potentially changeable'. I am not sure how to translate the text with Diels' punctuation. (JOU)

146. 201b4-5.

147. 409,27-32. The two quotations are slightly different: in 437,2 we read 'and became actual' (*kai egeneto energeiai*) which is missing from the quotation in 409,28, and we have a *hê* in 437,5 instead of the *kai* in 409,31.

148. 409,12-22.

149. Themistius, *in Phys.* 75,11-13 Schenkl. Here, in 437,12-13, Simplicius gives a fairly close paraphrase of Themistius' text.

150. Themistius, *in Phys.* 75,17-18.

151. I have not found any commentator who can be credited with such views. But the objection may have been well known for Philoponus also discusses it in his *in Phys.* 369,3-16.

152. Diels does not close the quotation anywhere. This seems to be a suitable place; Simplicius often comments on a quotation in this tentative way, e.g. at 443,30. (JOU)

153. Aristotle does not claim that what initiates change of place is not in place. For Simplicius, however, it is crucial because the chain of moving causes ends in the soul as a primary cause of all sorts of change. As an immaterial being, soul cannot be said to be in place in the strict sense.

154. Fragment 61 p. 33 Wehrli. Eudemus repeats Aristotle's view although we must remember that Aristotle allowed tactile perception to work differently.

155. Simplicius seems to take Andronicus of Rhodes, Peripatetic philosopher of the first century BC, to task for failing to realize that inner potentiality is a passive suitability (*epitêdeiotês*). At any rate, Andronicus' interpretation contradicts the Aristotelian doctrine (*Metaph.* 9.8) that the potential does not pass into actuality without the operation of something already actual. The actual must be external to the potential. As elsewhere in the commentary, Simplicius does not seem to respect Andronicus' reading. It may suggest that Andronicus' text was inferior. For further consequences of this attitude for Andronicus' edition of Aristotle's works, see J. Barnes, 'Roman Aristotle', J. Barnes & M. Griffin (eds.), *Philosophia Togata II*. Oxford 1997, 1-70, esp. 30. By contrast, H. Diels thinks that the reading is due to other commentators rather than to the transcript by Andronicus himself, see 'Zur Textgeschichte der Aristotelischen Physik', in *Abhandlungen der Kgl. Pr. Akademie der Wissenschaften zu Berlin*. 1882, Phil.-hist. Kl. I, 23, reprinted in W. Burkert (ed.), H. Diels, *Kleine Schriften*. Darmstadt 1969. For a similar report on Andronicus, see 450,17 ff.

156. Such are the dialectical arguments which Aristotle discusses in *Top.* 1.1, 100a29 ff.

157. This is the argument in *Phaedrus* 245C5-246A2.

158. This kind of argument is found in *Phaedo* 70D7-71A11.

159. *Phaedo* 70D7-71A11. This might have been a common opinion among Athenian Neoplatonists in those days, for Damascius, older contemporary of Simplicius and the last head of the Platonic school at Athens, also assumes that the argument from opposites proves only that the soul survives the body, not that it is immortal, see the excursus on the argument in his *in Phaedonem* 1 §§ 207-52 Westerink.

160. Reading *ousia* instead of *ousiai* in 441,14. (JOU)

161. Themistius, *in Phys.* 77,2-4 Schenkl.

162. Aristotle proves it in Book 8, 256a4 ff.

163. '*Tines*' with a paroxytone accent is interrogative and means 'which?' or 'what?'; with a barytone accent it means indefinitely 'some', as in the next sentence.

Alexander is proposing to transpose 'For what will be the two alterations of the one thing and into one form' at 202a35 to after 'but it is impossible'. With a barytone accentuation that sentence translates as 'for there will be some two alterations etc.', which Aristotle will then say is impossible. In the received version a question is asked, but not directly answered by 'it is impossible'. Alexander's proposal is purely stylistic with no consequences for the argument. (JOU)

164. See 202a26-7.

165. The iris contracts in the light, expands in the dark.

166. Aristotle returns to the subject in 447,14 ff.

167. 202a7. This is Aristotle's second definition.

168. The principle, shared by both Aristotle and Simplicius, that change is in the changed with the changer present and active must be modified if we are about to account for the motion of projectiles. In this case, the original thrower need not be present and active *always* nor is his activity *within* the projectile. In looking for a further cause (*Phys.* 8.10, 267a2-12) Aristotle assumes that it must be in external contact with the projectile. The thrower implants in the pockets of air behind the javelin the power to push it onwards. Simplicius objects by asking why we say that the motion is impressed on the air, rather than on the javelin. For, he rightly supposes, in this way we make the air a kind of unmoved mover, see *in Phys.* 1349,26-9. See R. Sorabji, 'John Philoponus', in R. Sorabji (ed.), *Philoponus and the Rejection of Aristotelian Science*. London 1987, 8.

169. The wax-model of knowledge was borrowed by the Stoics from Plato's *Theaetetus*. But their theory was based on the assumption that soul is a kind of subtle matter, which makes the account of acquiring knowledge in terms of physical constituents acceptable. See, e.g., *SVF* 1.484 (Cleanthes); 2.56 (Chrysippus). Simplicius may use it as a general model of sense-perception. In this he may have hoped to get support from Aristotle as well, who also referred to a certain wax-model of sense-perception, see *DA* 2.12, 424a17-18.

170. 202a13.

171. Alexander uses the adverbial form *apotetmêmenôs* ('as cut off') instead of the feminine participle *apotetmêmenê* ('cut off'). The reason for that may have been to emphasize that teaching is not to be taken as something isolated, although it might work in an isolated way.

172. 202a18-20.

173. Simplicius reverses the order between 'to be learning' and 'to be teaching' that we find in the text of the *Physics*.

174. The first definition is in 201b4-5, the second in 202a7.

174a. Simplicius thinks that Aristotle gave a third definition of change. This definition was reached by the analysis of change of quality. It says that change is the actualization of the potentially productive and passive as such (449,29-32). For Simplicius, this definition has several advantages. First, it links agency to being acted on. Second, since it has been demonstrated that change is completed from the activity of both the changer and the changed, it is reasonable that the new definition should include the activity of the changer as well. But here we have to face a problem. Simplicius asks why the definition refers to the activity of the potentially productive. For the potentially productive does not yet initiate change, whereas the changer is in actuality. In general, the potentially active is changed rather than initiating change. Simplicius surmises that by 'of potentially productive' Aristotle referred to a change which is communicated by the unchanging causes in relation to both the natural and the artificial change. 'For all changers initiate change in so far as they are in actuality, being unchanged in that respect' (450,5-6). Furthermore, if we mark out the cause which is unchanged in the strict sense, then we have to infer that such a cause is unchanged in all respects since it is actual unconditionally and exists forever. There is a necessity for the changer to communicate its form to the changed. The changer is itself changed because the initially dormant condition is aroused to the activity of initiating change. Simplicius develops this account by saying that things that thus initiate change are not changed insofar as they initiate change. For they do it insofar as they are perfect, but insofar as they are perfect they remain the same and are not changed. Thus Simplicius assumes that Aristotle gives three definitions of change, each being more and more adequate and exact. The last definition is also from Aristotle even if the text seems to be added in the book as from a marginal note and thus may seem muddled.

175. The movers of the spheres communicate form to the heavenly bodies in this way. They are not only exempt from generation and corruption but are of an immaterial nature as well. Simplicius briefly discusses this in his *in Cael.* 105,25-30.

176. Aristotle defines nature as inner source of change and rest in *Phys.* 2.1, 192b20-3, 193b3-4. But in 8.4, 255b24-256a3 Aristotle adds that the inner nature needs to be moved (*paskhein*) by an external cause. It is the nature of steam to rise, for example, but its rising presupposes that something generates the steam and something releases it. Andronicus here and in 440,14 ff. emphasises that the inner potentiality, as opposed to the external cause.

177. This must have been a usual procedure by that time, see H. Diels, 'Zur Textgeschichte der Aristotelischen Physik', in *Abhandlungen der Kgl. Pr. Akademie der Wissenschaften zu Berlin*. 1882, Phil.-hist. Kl. I, 31-2. He refers to Simplicius' *in Cat.* 88,25-7; 228,1-3.

178. Here Simplicius echoes Plotinus (*Enn.* 6.1-3) who also said that the supreme genera of being (change is one among them) characterize intelligible reality, while the Aristotelian categories apply at best, and only imperfectly, to the physical world. That is, the two schemes are complementary.

179. The reference is to Plato's concept of Ideas which, in one sense, are the perfect embodiments of the features they carry. On the other hand, Simplicius' distinction may recall the distinction we find at the beginning of the *Physics* (1.1) where we are told that physicists should start from the more obvious and known to us and proceed to the more obvious and known by nature.

180. *Cael.* 1.1, 268a1-3.

181. *Phys.* 4.11, 219a14-b8.

182. Deleted by Diels as a marginal gloss. (JOU)

183. *Cat.* 8, 9a28-10a10.

184. *Cael.* 1.1, 268a1-3, quoted in 451,22-3. Simplicius confirms this view in his in *Cael.* 7,8-10.

185. Diels prints this as a quotation, but it is a paraphrase. (JOU)

186. *An. Post.* 2.1, 89b23 ff.

187. On the Pythagoreans, see Aristotle, *Metaph.* 1.5, 986a24, 1.6, 987b27-9. In Plato, the reference may either be to the unwritten doctrines which say that the material principle is indefinite and thus infinite ('great and small'), or to the *Philebus* where Plato is talking about unlimited as a kind of principle.

188. On Anaximenes, see, e.g., A 5 D-K (Simplicius, *in Phys.* 24,26-25,1), A 6 (Ps.-Plutarch, *Stromata* 3), A 7 (Aëtius 1.3.26).

189. Thales, A 13 D-K (Simplicius, *in Phys.* 23,21; 458,23). Simplicius mentions this at *in Phys.* 24,13-16 as well.

190. Simplicius mentions in many passages that for Anaximander the unlimited is the intermediate, see 36,14; 14,12-18 – here we are told that the Peripatetic

Nicolaus of Damascus accused Diogenes of Apollonia of holding such a view; 458,25-459,1; 465,5; 484,12; 1266,37. In 465,14-15 we find the report that Anaximander put the *apeiron* between fire and air. On the *apeiron* as principle, see Anaximander B 1 D-K. Aristotle also repeats it in *De Gen. et Corr.* 2.5, 332a21-2; *Cael.* 3.4, 303b12-13. In his commentary to the passage in *GC*, H.H. Joachim defends the attribution of this view to Anaxagoras, see *Aristotle: On Coming to Be and Passing Away (De generatione et corruptione)*. A revised text with introduction and commentary by Harold H. Joachim. Oxford 1922, 224-5.

191. See also 27,5 ff.; 155,17; 459,13-17; 460,5; 532,12; 1069,24; 1123,21; 1184,19-22. In 166,10-11, Simplicius notes that homoiomeries (homogeneous stuffs) can be unlimited in form as well.

192. For Democritus, see e.g., Simplicius *in Cael.*, 294,33, with reference to Aristotle's *On Democritus* (fr. 208 Rose).

193. This implies that the Pythagoreans, just like the followers of Plato, made a distinction between numbers in physical things and transcendent numbers which are responsible for the mathematical order in the physical world. The quotation is from the Sacred Discourse and fr. 9 in *Iamblichus. De vita Pythagorica; accedit epimetrum De Pythagorae Aureo carmine* (ed.) A. Nauck, St. Petersburg 1884. See also 1102,20.

194. The testimonies on Hippasus' doctrines are few and this one is not included in D-K. Neither is any testimony of a similar content to be found there. If Simplicius is reliable here we can say that the fire, principle of the universe that is limited and moving eternally, produces the world according to the transcendent numbers as paradigms. Simplicius' wording recalls the *Timaeus* which in turn is said to encapsulate Pythagorean material.

195. *Timaeus Locrus* 3, 95B. The work is now considered to be a Pythagorean treatise from the Hellenistic period.

196. *Timaeus* 32C6-8. The passage 453,19-455,14 is listed as C 12 in M. Isnardi Parente, op. cit. in note 132.

197. There is no direct evidence that Plato placed the indefinite dyad among the intelligibles. But Plotinus, when discussing the structure of the intelligible realm and positing an intelligible matter, identified it as the substrate of the intellect. See *Enn.* 5.1.5,14-16; 5.4.2. The mention of place in 453,22-3 may also be a reference to the *Phaedrus* 247C.

198. In 161,6-19 Simplicius refers to Alexander of Aphrodisias as the source for his knowledge of Plato's unwritten doctrines, the lectures *On the Good*. As for Heracleides of Pontus, this is fr. 7 in Wehrli's collection.

199. The reference is to Porphyry's commentary *on Plato's Philebus* and registered as *Test.* 74F Smith. The passages in *Philebus* are 24A-27A discussing the nature of the unlimited. For Porphyry, it is not to be equated with the more – less altogether, see 25A.

200. *Philebus* 14D3-15D1, 24A1-25E2.

201. See Aristotle, *De Bono* fr. 2 Ross. In Alexander, see *in Metaph.* 55,20-55,35; 59,20-60,2 Hayduck. For the numerous references in Alexander's works to Plato's dialogues, see R.W. Sharples, 'The school of Alexander?', in R. Sorabji (ed.), *Aristotle Transformed*. London 1990, 83-113, esp. 91-2.

202. A surface was said to be the result of the sideways motion of a line and a solid from the upward or downward motion of a surface. (JOU)

203. Since number was conceived of as a plurality. Progress in arithmetic and algebra, as opposed to geometry, only became possible when 0 and 1 were treated as integers among the rest. (JOU)

158 *Notes to pages 74-77*

204. One source is Aristotle, *Metaph.* 13.7, 1081a21, 8, 1083b2, 9, 1086a5; *DA* 1.2, 404b25-7.

205. A variety of interpretations of this difficult phrase, in addition to those discussed by Simplicius, can be found in the notes ad loc. in Ross' edition. But doubtless Aristotle is referring to the fact that if successive odd numbers are added to one we get a succession of square, and in that way identical numbers, or, put geometrically, if successive odd numbers of unit squares are added to a given square we get a succession of square figures, whereas the addition of successive even numbers produces a series of numbers or figures with a great variety of properties or shapes. In geometry a gnomon is an L shaped figure partially surrounding another figure, normally a square. (JOU)

206. Half of twenty is ten which is even; half of ten is five, which is odd; half of two is one. (JOU)

207. A gnomon was a carpenter's square, used for ensuring that parts of a construction were at accurate right angles to each other. In Pythagorean diagrams the gnomon was some L-shaped area surrounding squares to form a larger square as in the diagrams at 457,18 and 457,22. The use in geometry is very similar and can best be illustrated by the diagram used by Euclid in *Prop.* 1.43.

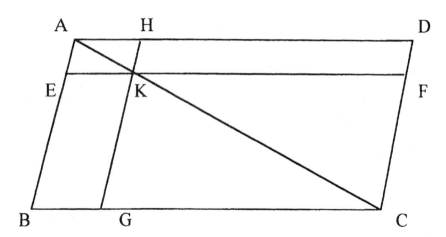

Here the gnomon is the area ADFKGB surrounding the parallelogram KFCG which in the theorem is said to be similar to the original parallelogram ABCD. EKBG and HDFK are the two complements and AHKE is what Simplicius calls 'one of the parallelograms about the diagonal'. The three together make the L shape like that of the Pythagorean gnomon which surrounds the parallelogram KFCG. (JOU)

208. The text reads *kata to hexês peritton arithmôn*, which Diels obelizes, suggesting *kata to hexês tôn perittôn arithmôn*. In any case the meaning is not in doubt. (JOU)

209. With regard to both odd and even numbers and with regard to both arithmetical and geometrical interpretations. (JOU)

210. That is, odd numbers were called gnomons because they produced square numbers $(1+3+5+ \ldots (2n-1) = n^2)$. Moreover, we must bear in mind that *pleura*, translated as 'side', in the language of Greek mathematics means 'factor'. For a variety of interpretations of Aristotle's passage on the Pythagoreans, see W.D. Ross' commentary on Aristotle's *Physics*, pp. 452-5.

211. Reading *tautais* instead of *toutois* in 458,16. (JOU)

212. Here Simplicius repeats the report he gives in 452,30-2, see notes 187-91.

213. See 453,1-4 with notes 191-2.

214. See Empedocles B6, B7, B22 D-K. He also adds Love and Strife as principal forces; Simplicius also makes mention of that in 459,12-13; 465,6-7.12.

215. Eudemus fr. 62 Wehrli. Eudemus thinks, and Simplicius agrees, that an infinity of ingredients with magnitude and in contact with one another will make an infinitely large whole (see also 462,3-5). But, on his view, a thing can contain only a finite number of ingredients which exist actually. Infinite divisibility is another matter, as it is potential only.

216. 203a21 and *DA* 1.1,404a4.

217. 458,26-7.

218. For Anaxagoras, see B1, B4a-c, B8, B11 D-K.

219. These ideas were common. The view that *creatio ex nihilo* is impossible had been generally accepted since Parmenides, while the notion that each thing is affected by what is similar to it, also has Presocratic roots which are reported in Theophrastus' *De Sensibus*.

220. Simplicius discusses Anaxagoras' theory of mixture at length in 485,21-7; 486,21-32; 487,11-18 and mentions it in 154,10-11.

221. Anaxagoras, B1 D-K. On the mixture in Anaxagoras, see also Simplicius' remarks in 485,21 ff.; 486,21-32; 487,11-18; 154,10; 177,20-178,7.

222. This principle contradicts Aristotle's remark in *Phys.* 1.1 and *An. Post.* 2.8, that the first things we encounter are wholes which are still undifferentiated.

223. Simplicius discusses this intellect in 34,18 ff.; 176,17-177,17 with reference to *Phaedo* 97B where Plato criticizes Anaxagoras' views on the intellect.

224. For a fuller quotation, see 300,31-301,4 (= B13 D-K). See also 35,14-18; 157,7-10; 164,24-165,7.

225. Simplicius offers this interpretation in 34,17-35,21, without mentioning the term 'demiurgic intellect'.

226. See Hippocrates(?), *De alimentatione* 23.

227. Anaxagoras B1 D-K.

228. This argument of Simplicius seems to presuppose an Aristotelian universe with a centre to which bodies, such as earth, tend. But Democritus would not have accepted that. (JOU)

229. Simplicius reports it in 10,8-12,13.

230. That is the four explanatory factors are called principles as well. Aristotle also uses the term *arkhê* to mean cause, see *Metaph.* 5.1, 1013 ff.

231. Omitting *arkhên ekhei* at 463,33, which Diels added from the Aldine edition, but which is not in the codd. (JOU)

232. *Timaeus* 29D7-E1. Simplicius adds a *pote* before *aitian* ('cause') and its punctation is different from Burnet's text which gives a full stop before 'he was good'.

233. 203b11-12, see the next lemma in Simplicius' commentary. The clause may be an allusion to Anaximander, fr. 2 (as has been noted by Ross in his commentary, pp. 546-7, with reference to Diels' conjecture).

234. It is in fact at *Phaedrus* 245D2-3, 4-6. (JOU)

235. Reading the text as in Plato's received text (*Phaedrus* 245D). Diels' text reads: 'it would not come from a principle'. If we are to accept this as correct, it is clearly a slip by Simplicius, for it is absurd.

236. Anaxagoras, B 12 (ap. Simplicius, *in Phys.* 164,24-165,1; 156,13-157,4), B

13 (ap. Simplicius, *in Phys.* 300,31-301,1), B 14 (ap. Simplicius, *in Phys.* 157,7-9); Empedocles, B 17 (ap. Simplicius, *in Phys.* 158,13-159,4).

237. For necessity in Empedocles, see B 116, quoted in part in Simplicius 1184,9-10. It might refer to the inevitability of the most general of natural processes, the exchange of love and strife, as well as that of the elements.

238. On Anaximander's notion of unlimited as the intermediate, see 452,32 with n. 190.

239. The term 'suitability' (*epitêdeiotês*) is Neoplatonic, but Simplicius' remark is backed by Aristotle's claim that natural things have a principle of movement, not of producing it but of suffering (*paskhein*) on the occasion of an external cause. But they can only suffer movements which are proper to them (*Phys.* 8.4, 255b29-30). In commenting on that passage, Simplicius again uses the term 'suitability' in 1218,8.

240. Reading *tinos* at 465,30-1 instead of *titos* which is clearly a misprint. (JOU)

241. Eudemus fr. 63 Wehrli. Eudemus starts from the Aristotelian definition of continuity as infinite divisibility (200b20).

242. A literal translation; 'never coincides with' is geometrically more accurate.

243. Reference to Anaximander A 14 D-K (= *Placita* 1.3.3).

244. For Epicurus, see also Diogenes Laertios 10.42. This is *Test.* 297 in Usener's *Epicurea* (Leipzig 1887). He refers to Alexander's *Quaestiones* 3.12 for a parallel. For similar arguments, see Epicurus, *Letter to Herodotus* 41, and Lucretius 1.958-67.

245. See also Democritus A 67 (ap. Simplicius, *in Cael.* 295,1-10).

246. Eudemus fr. 64 Wehrli. Archytas' argument was elaborated in later times, for which see R. Sorabji, *Matter, Space & Motion*. London-Ithaca NY 1988, 125-8. Alexander *Quaest.* 3.13, 106,37-107,4 exploits Aristotle's view that things have place, the two-dimensional inner surface of the thing's surroundings. The universe has no surroundings and thus there is no place for the hand to stretch into. Simplicius repeats Alexander at *in Cael.* 285,21-7. Here at 467,35-468,3 he gives the same solution, but adds that nothingness does not prevent the stretching: it *neither* prevents it *nor* accommodates it.

247. At 467,37 Diels has *hulên* – matter – which I take to be a misprint for *holên* – whole. (JOU)

But we cannot rule out either that the universe is conceived of as having matter as its place. It is possible if prime matter is equated with three-dimensional extension, see *in Phys.* 229,6; 230,19-31; 232,20-4; 537,13; 623,18-19. The reference might be to the *Timaeus*. In 643,5-12 Simplicius interprets the dialogue to claim that matter is the place of forms. (PL)

248. Eudemus fr. 65 Wehrli.

249. Many texts of Aristotle, including that of Ross in his edition read 'or'. But see 468,26; 'in number and in magnitude'.

250. Aristotle demonstrates it in Book 8 of the *Physics*. As for Plato, Simplicius may refer to the interpretation of the *Timaeus* according to which the creation myth is not to be taken literally. The majority of Platonists (the most notable exception were Plutarch and Atticus) took these passages as saying that the cosmos was not created and would not perish. The dispute over the eternity of the world marked the dividing line between pagan and Christian philosophy in the age of Simplicius (and before). Proclus listed eighteen arguments against the creation of the world, which were responded to by John Philoponus in his *De aeternitate mundi contra Proclum*.

251. Possibly delete *ê* in 470,11 and read 'of aquatic animals'. (JOU)

252. The basic meaning of this word is 'sling'. *LSJ* defines it with regard to a ring as 'hoop of a ring in which the stone was set as in a sling, esp. the outer or broader part round the stone, collet'. But we seem to need a sense here in which the sling in some way breaks the continuity of the ring. (JOU)

253. See Pausanias, *Description of Greece*, 4.18.4 ff.

254. Aristomenes was a Messenian leader in the second Messenian war in the seventh century BC. For a vivid account of his escape with the fox's aid see Pausanias, *Description of Greece*, IV, 16, 6-7. (JOU)

255. 'Reality' translates *huparkhon*. The term and its cognate *huparxis* signify a special kind of existence. In the late Neoplatonists at Athens it meant independent existence or divine being. It can be contrasted with *hupostasis*, a more vague term which signifies a general way of existence. For the distinction in Proclus and some later Neoplatonists, see C. Steel, '*Huparchô* chez Proclus', and D.P. Taormina, 'Anima e realtà del conoscere. *Hupostasis* e *Huparxis* nei Commentari tardoantichi al *De anima*', both in F. Romano & D.P. Taormina, *HYPARXIS e HYPOSTASIS nel neoplatonismo*. Firenze 1994, 79-101, 101-31.

256. 'Attribute' translates *sumbebêkos* without a preceding preposition; here we have *kata sumbebêkos*, which normally indicates that the attribute referred to is not necessary but, in the technical sense, an accident. As will be seen, commentators disagree on the significance of the change. The translation 'attributive' is intentionally neutral. (JOU)

257. For a report on previous views on the unlimited, see 452,21-455,14. There Simplicius interprets Anaximander's *apeiron* as a kind of intermediate, see also 443,36-444,1; 458,25-35; 459,3.

258. 204a29-30.

259. It appears from 474,2-5 that Simplicius used a text reading *ouk endekhetai ousian einai to apeiron hôs energeiai kai hôs ousian kai arkhên*; the received text does not contain the first *ousian*. I have translated this as probable.

260. Diels marks a lacuna at 473,35, the quotation from Aristotle being missing. (JOU)

261. The word *ousia* is an abstract noun etymologically and semantically connected with the verb *einai*, which simply means 'to exist'. To apply it only to substances is to use it technically. (JOU)
The distinction mirrors the distinction of *per se* and derivative existence and thus can refer to the distinction between *hupostasis* and *huparxis*. (PL)

262. 'This' (the individual thing) and 'to be this' (the essence – *to ti ên einai* – of the individual thing) can be equivalent only if the 'this' signifies a form. In that case, we can say that the form is equal to its essence, its definiendum. As soul is a form, the identification of a particular form with its essence is possible.

263. Eudemus fr. 66 Wehrli.

264. Simplicius has *diairousi* at 204a34; the received text of Aristotle has *merizousin*. (JOU)

265. For Diogenes of Apollonia, see 452,32; 458,25; for the Pythagoreans, see 452,29; 453,5; 455,12-20; 476,20 ff.

266. This was mentioned in 456,1 ff.

267. But at 475,29 Simplicius has *kai en tois mêden ekhousi megethos*; if this is an exact quotation from his text of Aristotle we should translate the passage as 'objects of intellect and things that are without magnitude'.

268. Aristotle uses the word *noêtos* to mean object of the intellect. For Simplicius, on the other hand, it also refers to a separate realm of being, that of the intelligible entities (e.g. the Ideas and the Intellect). Therefore, in a Neoplatonic context the term can also be rendered as 'intelligible'.

269. Reading *epistêmonos* instead of *epistêmês* at 475,31 with Vitelli. Cf. 469,24 and 510,17. (JOU)

270. *Phaedrus* 247D5-E1. There are two significant gaps in Simplicius' quotation. The received text of the *Phaedrus* contains an *autên* before 'justice' and a *pou* before *hetera* ('something').

271. *logikôs*. For this translation see 476,28 below. (JOU)

272. *Top.* 1.1, 100a18-21.

273. 204b10.

274. For Proclus, imagination is the place of mathematical objects. As he says in the second introduction to his commentary on Euclid's *Elements*, *logoi* (concepts) in discursive reason will be projected into imagination where they become extended and thus accessible to geometrical inquiry. Simplicius was doubtless familiar with the doctrine, see *in Phys.* 512,19-26; 621,33; 623,15-16.

275. 205a29.

276. 479,300 ff.

277. 187a12 ff.

278. 205a29.

279. The received text of Aristotle, and that possessed by Simplicius, read 'air is cold, water is moist'; as will be seen, Simplicius believes this to be a scribal error, since Aristotle in *De Gen. et Corr.* at 331a4-5 says that air is primarily moist, water primarily cold, air not cold but warm. I have translated the text as conjectured by Simplicius. (JOU)

280. Eudemus fr. 67 Wehrli. The Empedoclean elements are fire, air, water and earth.

281. See Heraclitus A 10, B 30, 64, 66, 90 D-K.

282. The Stoic notion of conflagration (*ekpurôsis*) draws on the doctrine of Heraclitus (see B 30 and 66 D-K). Simplicius repeats this kinship at his *in Cael.* 94,4-6.

283. It is parallel to Heraclitus B 66 D-K.

284. 188a30 ff.

285. On listing the basic qualities of the four elements Aristotle assigns wet and warm to the air in *GC* 2.3, 330b4, 22; 331a5.

286. Simplicius has *einai apeiron* in the commentary. The MSS of Aristotle vary among themselves. (JOU)

287. The text of Aristotle is doubtful in parts of this passage and the MSS vary. In addition various editors wish to transpose parts of a not very well organized argument. It is not always clear exactly what text Simplicius had before him. (JOU)

288. *Cael.* 1.7, 274b26-8. Simplicius comments on it in his *in Cael.* 230,21-33.

289. Reading *prolambanei* with Vitelli at 482,13 instead of *proslambanei*. Cf. 482,28. (JOU)

290. In Book IV Aristotle maintains that there is nothing beyond the sphere of the fixed stars; therefore it has no container and consequently is not in a place. Simplicius in his corollary on place (*in Phys.* 601-45) argues strongly against this view of Aristotle. (JOU)

291. This argument presupposes the doctrine of natural places for each of the four elements. As things are, the natural place of earth is at the bottom, to which the clod will travel and where it will remain. But if earth is everywhere the clod will have no reason to travel to the bottom or anywhere else. (JOU)

292. 4.7, 214a16-17. The demonstration that void does not exist runs until 217b29.

293. 1.4, 187a26-188a17.

294. The final three sentences of the lemma given here are omitted in the text. But they are referred to at 486,27 in the commentary. (JOU)

295. On natural place, see 4.1, 208b11; *Cael.* 4.3, 310b3.

296. Simplicius approves Aristotle's criticism of Anaxagoras' theory that the unlimited is in itself. Anaxagoras may have come to this conclusion from the premises that there is nothing which could be bigger than the unlimited and thus there is nothing which could contain it. He may have added that where a thing is, it is due to its nature to be there. The unlimited is in itself, therefore it by nature supports itself to be in itself. Besides Aristotle's report, we do not have evidence for this view in Anaxagoras. For self-supporting, see also 601,12.

297. Reading *kata* instead of *para* at 486,30; cf. 487,12. (JOU)

298. *Phys.* 4.8, 215a1-6.

299. *Phys.* 4.2, 209a31 ff.

300. 461,11, see also 1186,30-4. In 34,18-35,21 we find a threefold distinction attributed to Anaxagoras; he is alleged to divide the intelligible into a unified and a differentiated realm.

301. See *Phys.* 208b10 ff. and Simplicius ad loc. 525.21 ff. (JOU)

302. See *Cael.* 4.1, 308a14 ff. with Simplicius' comments at his *in Cael.*, 678,20-682,3.

303. This is the only evidence for the existence of Zenobius the Epicurean, see K. von Fritz's entry 'Zenobios 3' in *RE* II 19/2, col. 12. Neither is there any other evidence that Alexander composed a work against Zenobius. There is a possibility, on the other hand, to explain *antigegrammenois* – probably hinting at a missing *logois* – as referring to arguments. This would allow us to say that these arguments were included in Alexander's commentary on Aristotle's *Physics*. The commentary is now lost but was a principal source for Simplicius.

304. 'General truths' translates *koina*, referring to common notions (see *koinê hupolêpsis* in 9,33). As he is discussing a physical theory, Simplicius is content with remaining within the Aristotelian framework here. But elsewhere, in 132,18-19, he takes the opportunity to distinguish two kinds of genus, common (*koinon*) and substantial (*ousiôdês*), the latter signifying Platonic forms. See also 490,32-491,1.

305. Eudemus fr. 68 Wehrli.

306. Literally: better to give *poson* an oxytone accent in both 'to be a quantity' and 'will be some quantity', and not to give it a barytone in 'for the unlimited to be a quantity' as Alexander wishes. But the definite, which the barytone indicates, should be understood in 'it will be some quantity'. (JOU)

307. In 490,31 reading *tode ti*. Diels has *to de ti*. (JOU)

308. For a similar remark, see 490,3-6.

309. This is the expression regularly used in the account of time in Book 4. Time is that by which we can reckon how much change has occurred, and the amount of time that has elapsed is measured by change. Thus three days is the number of the changes in the motions of the sun round the earth. (JOU)

310. It will be proved in 208a5 ff.

311. This book is the *De lineis insecabilibus*, now generally held to be spurious. Diogenes Laertius (5.1.42) also attributes it to Theophrastus.

312. Euclid, *Elements* 6.10, see also 511,22-3.

313. Eudemus fr. 69 Wehrli.

314. This may be a reference to Plato's lectures *On the Good* which is one of his unwritten doctrines. Simplicius' main, if not only, source for the application the doctrine to physics may have been Alexander of Aphrodisias, see also 453,28; 454,18.

315. Simplicius' solution is based on the eternity of the universe, otherwise the actual cuts cannot go on without limit. On the other hand, by positing an actual infinity by division, he gets close to Zeno's notion of finite spatial distance as being succesively bisected (see *Phys.* 6.2, 233a24-8)

316. Reading *aitian* instead of *aition* at 494,22 because of *tên heteran* at 494,25. Cf. *tên aitian tou apeirou* at 495,16. (JOU)

317. The reference is to 206a33. Philoponus has the same version as Simplicius.

318. For Aristotle's *brakhulogia* which deceives many of his commentators, see also 112,30.

319. *sumphainei*; the verb is listed in *LSJ* only in the passive. The Aldine edition reads *sunuphainei*.

320. Philoponus gives the latter version (see his *in Phys.* 469,21-470,27) and explains it also in terms of cutting in a definite ratio.

321. But in his independent *Corollary on place, in Phys.* 601-45 he raises severe problems about the doctrine of the limited universe which is here left unchallenged. *De Cael.* 1.5-7. (JOU)

322. 494,4-11.

323. Simplicius refers to this doctrine in 454,22 as well. For an examination of the passage, see M. Isnardi Parente, op. cit. in note 132. She lists it as C 13 in her collection of the testimonies on Plato's unwritten doctrines.

324. See the note to 499,35. (JOU)

325. But as far as Plato's doctrine, or it may be better to talk about the doctrine of the Old Academy (and Xenocrates is the primary figure in this respect), we have reports saying that mathematical number is identified with ideal number or idea-number, see Asclepius, *in Metaph.* 397,19 (fr. 104 Isnardi Parente) and there is an unnamed reference in Aristotle, *Metaph.* 7.2, 1028b24-6 (fr. 103 I.P.)

326. 'Secondary' and 'tertiary' are *LSJ*'s not very revealing translations of the rare words *deuterôidoumenas* and *tritôidoumenas*, the meaning of which is unclear. Etymologically they seem to mean something like 'making a second (third) journey'. They should probably be understood with reference to 499,15-20 above. Given the monad and the numbers from 2 to 10, the decad, we get composites of numbers from the successive numbers and the decad. Thus twenty is two decads, thirty three decads and so on, getting as far as 100, which is ten decads, by the second journey through the numbers. Then 200 is a decad of twenties, 300 is a decad of thirties, and so on to a thousand by the third journey. A fourth run through decads of thousands, and so on, would be needed for still higher combinations of the numbers in the decad. This seems to be what is referred to at 499,18-20 above. (JOU)

327. 207a6.

328. This final sentence seems to contradict what precedes it.

329. For Melissus, see B 2 (in Simplicius' *in Phys.* 29,22 ff., 109,20 ff.), B 3 (Simplicius, *in Phys.* 109,31).

330. Parmenides B 8.44 (see in Simplicius, *in Phys.* 146,17).

331. The proverb about 'joining flax with flax' quoted by Aristotle in the lemma under discussion. (JOU)

332. See 225,21; 226,5 ff.; 233,14; 249,25 ff.; 542,21-2.

333. On Simplicius' discussion of Plato's lectures *On the Good*, see 451,18; 453,28; 542,12; 545,24. 'Fluid' is a Platonic term to describe matter. The concep-

Notes to pages 127-132 165

tion, if not the term, may have originated in Heraclitus, but was employed mainly by Platonic authors, see, e.g. Numenius fr. 52 Des Places (in Calcidius, *in Tim.* 296); Nicomachus of Gerasa, *Introduction to Arithmetics* 1.1.3. They may have hoped to find support in Plato's *Timaeus* as well (43A5-6). Simplicius' older contemporary, Damascius, also refers to the theory and ascribes it to Aristotle (*De Principiis* 2.172, 16-22 = Aristotle, fr. 207 Rose).

334. The reason why Simplicius takes Aristotle to say that the great and small is present among the intelligibles may be that he is following Alexander who reports that for Plato the principles of all things, including the Ideas themselves, are the one and the indefinite dyad which is also called great and small. Alexander refers to Aristotle as the source of this information, see Simplicius, *in Phys.* 151,6-19.

335. Simplicius refers to the intelligible matter (*hulê noêtê*). Although he does not use the term in the commentary, he may have been well versed in the theory through Syrianus (*in Metaph.* 186,32) and Plotinus (in general 2.4.2-5; see also 2.4.5, 16; 5.1.3, 23.)

336. i.e. addition at one place of parts taken away at another place. (JOU)

337. Reading *to hen men* at 207b2 with MSS E and I. See 504,29-30 below.

338. It appears from 505,29-30 that S. omits *tês dikhotomias*, which follows in most codd. of Aristotle. (JOU)

It appears from 506,21 that Simplicius read *all' aei gignetai* at 207b14 with MSS VPS. (PL)

339. This and the following sentences contain reference to the doctrine of *henades* ('units') as we find it in Proclus. According to this doctrine, units are the final ingredients of everything divisible, while themselves being exempt from all kinds of division. The reason for their being indivisible is that they do not contain any material element, physical or intelligible. For a full exposition of the theory, see Proclus, *El. Theol.* §§ 6; 64; 113-15; 135-9; 149. For a recent analysis, see L. Siorvanes, *Proclus: Neo-Platonic Philosophy and Science*. New Haven 1996, 167-75. Here Simplicius uses the doctrine to underline that further division must stop at one point. The item preventing further division must be indivisible altogether. Therefore it must be entirely formal.

340. Diels printed *duo kai tria*, but in his apparatus says that his sources all read the numbers in the reverse order. He says that Aristotle's text reads as in his text, but in his addenda on p. 1463 he noted that the best MSS of Aristotle read *tria kai duo* and that he wished to correct his text accordingly. (JOU)

341. The meaning of 'paronymous' in Aristotle's works is not always clear, and its use here does not reflect the definition given in *Cat.* 1a13. The distinction being made here is between the abstract, monadic or unitary number of pure mathematics and applied number such as twelve men. This number of men may be in a field, but abstract numbers are not anywhere. Both Aristotle and Simplicius think this difference crucial in giving an account of unlimited numbers. (JOU) See also Simplicius, *in Cat.* 86,29-33 (*grammatikos* is derived from *grammatikê*). (PL)

342. Time measures off the minutes and years that pass by and also we assign numbers to time such as '10.30 a.m.' or 'ten minutes'. (JOU) Aristotle writes about this in *Phys.* 4.14, 223a24-5. (PL)

343. *Phys.* 6.1, 231a29-b6.

344. For Plato, the cosmos is spherical, see *Timaeus* 33B4. Aristotle argues for the spatial finitude of the world in *Cael.* 1.5-7.

345. Simplicius derives *ouranos* (heaven) from *horan anô* (to look upwards). Cf. Plato, *Cratylus* 396b-c: *hê de au es to anô opsis kalôs ekhei touto to onoma kaleisthai, 'ourania', horôsa ta anô*. (JOU)

166 Notes to pages 132-136

346. The sphere of the fixed stars is everlasting and divine while the rest of the universe comes to be and passes away. Diels has *aidiô* at 507,33, without iota subscript, an obvious misprint. (JOU)

347. *Cael.* 1.5, 271b28-273a5. As for the notion that nothing lies outside the heaven, see *Phys.* 3.4, 203b28-30.

348. The suggestion is that the relevant increase of number is that of the fragments cut off in the dichotomy of a magnitude, not the increase of separate, i.e. abstract, number. (JOU)

349. Reading *aporia* with F. Diels has *apeiria* – limitlessness. (JOU)

350. In his edition Ross has *dioti* in 207b26; the translation of his text is 'and why every magnitude is divisible into magnitudes'. Simplicius, like the MS. E of Aristotle, has *hoti*, and that is here translated. (JOU)

351. cf. *Nicomachean Ethics* 1096,7-8: 'those who introduced this opinion did not posit Forms where they recognized a prior and a posterior.' The prior and posterior occur when one term is in some way contained or presupposed in another. Other cases dealt with by Aristotle as of this sort are souls, (in which the less complicated souls of plants are in a way contained in those of non-rational animals, and these souls in those of rational animals) and states. See *DA* 414b30ff. and *Pol.* 1275a34ff. (JOU)

In the Platonic tradition, an entity is called prior to another entity if it can exist without the posterior whereas the posterior cannot exist without the prior. It does not allow the existence of Ideas because the members of the series cannot be included into a common genus, see Aristotle, *EN* 1.4, 1096a17; *EE* 1.8, 1218a1; *Metaph.* 13.4, 1079a14; 14.1, 1087b23. See A.C. Lloyd, 'Genus, species and ordered series in Aristotle', *Phronesis* 7 (1962), 67-90. (PL)

352. See the definition of time in *Phys.* 4.14, 223a24-5.

353. Derivative naming from a central case (*pros hen, apo henos*) is typified by places, complexions, diets, etc. being called healthy from their relation to the central case of the healthy person. See *Metaph.* 1003a33 and 1030b30-3 and also *EN* 1096b27-9. (JOU)

Elsewhere, at his *in Cat.* 86,29-33, Simplicius considered it as a certain paronymy. In Aristotle, paronymy (*Cat.* 1a12) may be regarded as a kind of *pros-hen* relation. Simplicius knows that the Old Academy (most notably Speusippus) also used the term in this meaning, see *in Cat.* 29,5 ff., 36,25 ff., 38,21 ff. (PL)

354. The point is that the word *kinêsis*, regularly translated as 'change' in this discussion, here refers specifically to change of place; in its wider sense the word covers alteration and quantitative change as well as motion. (JOU)

355. *Rep.* 7, 527C5-6.

356. Reading *ou monon ou* at beginning of 511,15.

357. *Elements* 6.10. Simplicius referred to it in 492,6 as well.

358. The first of Euclid's theorems is; 'On a given limited straight line to construct an equilateral triangle.' The proof requires the drawing of two intersecting similar circles, the joint diameters of which are greater than the given straight line, the extremities of which are the centres of the circles.

AB is a given finite straight line; circles with radius AB intersect at C. ABC is the equilateral triangle. (JOU)

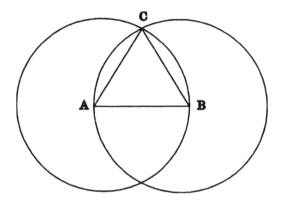

359. This argument of Alexander's well illustrates the notion of the unlimited as simply that which can be added to and extended and which has not a boundary preventing extension. Clearly his argument does nothing to suggest infinity in a modern sense. (JOU)

360. Reading *ti kai exôthen* with Diels at 512,16.

361. Simplicius shares Proclus' view that mathematical bodies reside in the thought and imagination of the geometer, see also *in Phys.* 621,33; 623,15-16.

362. See Democritus A 1 (*D.L.* 9.45), A 135 (Theophrastus, *De sensibus* 50), A 254 (Plutarch, *Adversus Colotem* 1110E). The atoms are not coloured. Colour and sweetness exist just by convention, in fact there are only atoms and void.

363. Simplicius has in mind proofs of the possibility of certain constructions by actually drawing such a construction, as in the proof of the possibility of constructing an equilateral triangle referred to above. (JOU)

364. *Phys.* 4.1, 208b22-5.

365. Simplicius may be referring to the Epicurean notion of *minima*, aspects of the atoms, see Epicurus, *Ep. Herod.* 56-9. But there is also a possibility that the reference is to some Platonists. Xenocrates developed a notion of atomic quanta of magnitudes, see frr. 127-47 Isnardi Parente. Aristotle may have been referring to the latter position when in *Cael.* 1.5, 271b10-2 he said that the postulation of a smallest magnitude would disturb mathematics. In *Cael.* 3.4, 303b20-1 he says that the acceptance of individual bodies is against the assumptions of mathematics.

366. Aristotle calls matter, form and privation principles of all physical things, see *Phys.* 1.7, 189b30-191a22.

367. A reference to the discussion in 542,21-453,18; 458,19-462,22.

368. *Proslêpsis*; possibly we should read *prolêpsis* – presupposition – and eliminate 'additional'. Cf. 483,13 and note ad loc. The inexhaustibility of generation seems to be a fundamental premiss of the argument. (JOU)

369. Alexander *Quaestiones* 3.12, 104,24-9, defends Aristotle's view, *Phys.* 3.8, 208a11-20, that what is limited does not have to be limited *by* something outside of it, in the way that what is touched has to be touched by something. Simplicius disagrees with Alexander at *in Phys.* 516,20-32. Further, at *in Cat.* 125,10-12 he does not subsume touching under the category of relation.

370. Simplicius repeats the definition of point at his *in Cat.* 136,29-30. The context suggests that he is acquainted with it through Iamblichus. Proclus also discusses it (*in Eucl.* 95,21-96,15) and says that the definition comes from the Pythagoreans.

371. I take the latter part of this lemma to read in the version used by Simplicius as *all'ou dia touto exô tês poleôs ti estin ê tou têlikoutou megethous ho ekhei* etc. See especially 517,20-2 below. The translation follows Simplicius' preferred interpretation stated in the commentary below. (JOU)

372. For a connection between mathematical magnitudes and imagination, see 621,33; 623,15-16. The expression 'beyond the heavens' (*exô tou ouranou*) may refer back to the report on Archytas in 467,26 ff.

373. Diares is the man whose son is taken by Aristotle to illustrate that individual substances as such are not what is apprehended by the senses essentially. They are incidental perceptibles, see *DA* 2.6, 418a21.

374. Eudemus fr. 70 Wehrli.

English-Greek Glossary

absurdity: *apemphasis, atopia, atopon*
account: *logos*
action: *poiêsis, to poiein*
activity: *energeia*
actuality: *energeia, entelekheia*
affection: *pathos*
affective: *pathêtikos*
agent: *to poioun, poiêtikos*
air: *aêr*
alteration: *alloiôsis*
analogy: *analogia*
analysis: *diarthrôsis*
animal: *zôon*
animate: *empsukhos*
apodictic: *apodeiktikos*
argument: *epikheirêma, epikheirêsis, logos, paradeixis*
to articulate: *diarthroun*
artist: *grapheus*
to ascend: *anabainein*
atom: *atomon*
attribute: *sumbebêkos*
axiom: *axiôma*

beginning: *arkhê*
being: *huparxis, ousia, to einai, to on*
body: *sôma*
breath: *pnoê*

category: *katêgoria*
cause: *aitia, aition*
ceasing to be: *phthora*
change: *kinêsis*
to characterize: *kharaktêrizein*
circle: *kuklos*
with circular orbit: *kuklophorêtikos*
to class: *tattein*
classification: *diataxis*
to co-exist: *sunhuparkhein*
colour: *khrôma*
coming to be: *genesis*
commensurate: *summetron*
comment: *hupomnêmatismos*

commentary: *hupomnêma*
completeness: *teleiotês*
compound: *sunthetos*
concept/conception: *ennoia*
conclusion: *sumperasma*
condition: *diathesis*
conditional proposition: *sunêmmenon*
confirmation: *pistis*
connective: *sundesmos*
consequence: *akolouthia*
constant: *ta hôsautôs ekhonta*
contact: *haphê, thixis*
to contain: *periekhein, sullambanein*
contiguous: *sunaphês*
continuous: *sunekhês*
contradictory: *antiphatikos*
contrary: *antikeimenon*
to convert: *antistrephein*
copy (of a manuscript): *antigraphos*
creative: *poiêtikos*
to cut: *temnein*
cut (off): *diairêma, tomê*
cycle: *periodos*

to define: *diorizein, horizein*
definite: *hôrismenos*
definition: *horismos, diorismos, logos*
demonstration: *apodeixis, deixis*
demonstrative: *apodeiktikôs*
dichotomy: *dikhotomia*
difference: *diaphora*
discrete: *diakekrimenon, diorismenon*
disease: *nosos*
dissimilarity: *anomoiotês*
disposition: *diathesis*
distinction: *diairetikon, diakrisis*
divisible: *diairetos*
division: *diairesis, diairetikon, diorismos, tomê*

earth: *gê*
element: *stoikheion*
end: *peras, teleutê*

essence: *ousia*
equivocal: *homônumos*
eternity: *aidion, aiôn*
existence: *hupostasis, ousia, to einai, to on*
explanation: *aitia, aition, exêgêsis, to dioti*
extreme: *akron, peras*

fact: *pragma, to hoti*
fire: *pur*
to fit: *epharmottein*
to follow: *akolouthein*
food: *trophê*
forced: *biaios*
form: *eidos, morphê*

general: *katholikos, katholou, koinos*
genus: *genos*
geometry: *geômetria*
gnomon: *gnômôn*
goal: *telos*
god: *theos*
to grasp: *lambanein*
the great and small: *to mega kai mikron*
growth: *auxêsis*

halt: *stasis*
health: *hugieia, to hugiainein*
HERE: *têde*
homoiomeries: *homoiomereis*
hot: *thermos*
house: *oikia*
hypothesis: *hupothesis*

ignorance: *agnoia*
immaterial: *aülos*
impossible: *adunatos*
inanimate: *apsukhos*
incidental: *kata sumbebêkos*
incomplete: *atelês*
incorporeal: *asômatos*
increase: *auxêsis, epitasis*
indefinite: *aoristos*
individual: *atomos*
indivisible: *atomos, adiairetos*
induction: *epagôgê*
inquiry: *theôria, zêtêsis*
instantaneous: *akhronos*
intellect: *nous*
intellective: *noeros*

intelligible: *noêtos*
intermediate: *to metaxu*

kind: *eidos*
knowledge: *epistêmê, gnôsis*

to lie: *to keisthai*
life: *zôê*
light: *phôs*
limit: *peras*
limitlessness: *apeiria, apeiron*
line: *grammê*
love: *philia*

magnitude: *megethos*
man: *anthrôpos*
material: *enulos, hulikos*
matter: *hulê*
to mean: *sêmainein*
measure: *metron*
method: *methodos*
mixture: *mixis*
moderate: *summetros*
motion: *kinêsis*
multiplicity: *plêthos*

name: *onoma*
nature: *phusis*
necessary: *anankaios*
necessity: *anankê*
negation: *apophasis*
not-being: *to mê on*
number: *arithmos*

occasion: *kairos*
occurrence: *hupostasis*
the one: *hen*
opinion: *doxa*
opposed: *antikeimenon, enantios*
opposite: *antikeimenon*
order: *taxis*
organ: *organon*
otherness: *heterotês*
own: *idios, oikeios*

part: *meros, morion*
participation: *methexis*
particular: *kathekastos, merikos, tode*
passage: *lexis*
passive: *pathêtikos*
perception: *aisthêsis*
to perish: *phtheirein*

to persist: *hupomenein*
place: *topos, to pou*
point: *sêmeion, stigmê*
position: *thesis*
posture: *to keisthai*
potentiality: *dunamis*
power: *dunamis*
premiss: *protasis*
primary: *prôtôs*
principle: *protasis*
privation: *sterêsis*
problem: *aporia, problêma*
process: *proodos*
product: *apotelesma, poiêma*
proof: *apodeixis, deixis*
to prove: *apodeiknunai, deiknunai*

quality: *poiotês, to poion*
quantity: *posotês, to poson*

rational: *logikos*
ratio: *logos*
reason: *aition*
to receive: *dekhesthai*
to refute: *dielenkhein, elenkhein*
relation: *pros ti, skhesis*
rest: *êremia, monê*
rod: *rhabdos*
rotation: *strophê*

science: *epistêmê*
seed: *sperma*
self-changing: *autokinêtos*
self-identity: *tautotês*
self-sufficient: *autotelês*
sense: *sêmainomenon*
separate: *khôristos*
separation: *diakrisis, exanastasis*
to show: *deiknunai, dêloun*
simple: *haplos*
to solve: *luein*
soul: *psukhê*
species: *eidos*

statue: *andrias*
straight: *euthus*
to stretch out: *ekteinein*
study: *theôria*
subject-matter: *hupokeimenon*
substance: *ousia*
substrate: *hupokeimenon*
suitability: *epitêdeiotês*
surface: *epipedon, epiphaneia*
surrounding: *perithesis*
syllogism: *sullogismos*

tangible: *haptos*
temporal: *enkhronos*
tendency: *rhopê*
THERE: *ekei*
thing: *pragma, to on*
thought: *noêsis*
time: *khronos, to pote*
to touch: *haptesthai*
transformation: *metabolê*
to transit: *metabainein*
travel: *phora*
to traverse: *diexerkhesthai*
triangle: *trigônon*
true: *alêthês*

unchanged: *akinêtos*
uniformity: *homalotês*
unit: *hen, henas monas*
unity: *henôsis*
universal: *katholou*
universe: *kosmos, to pan*
unlimited: *apeiron*

vital: *zôtikos*
void: *kenon*

water: *hudôr*
way: *tropos*
weight: *baros, barutês*
whole: *holos*
word: *logos, onoma*

Greek-English Index

adêlos, obscure, 437,30; 449,24
adiairetos, indivisible, 452,3.5; 454,4; 456,15; 471,22.24.27.28; 473,13.14.20.23; 475,3; 476,14.19.23; 499,13; 504,35; 505,2.5.9.10.12.20.22; 515,24
adialeiptos, unending, 454,6
adiaspastos, indestructible, 406,7
adiastatos, indivisible, 487,26; (*to adiastaton*), indivisibility, 514,30
adiexitêtos, without exit, 470,17; cannot be traversed, 470,27; 471,15; 471,31; 474,16; inexhaustible, 510,20; 513,24
adiexodos, without exit, 470,30
adiorthôs, incorrect, 500,7
adunatos, impossible, 394,9; 427,17; 442,10; 443,2.7.9.12.15.16.23.24.26.29; 446,29; 467,23; 470,1; 473,23; 474,24; 475,15; 480,30 etc.
aeikinêtos, always changing, 435,3.25
aêr, air, 399,1.13; 410,8; 423,19; 426,1.3.4.7; 452,31; 453,21 etc.
agathos, good, 429,9; 464,6; 503,12
agathotês, goodness, 464,4
agenêtos, ungenerated, 398,5; uncreated, 464,17.22.24.26.29.30; 465,3.14
agnoein, be ignorant, 394,10
agnoia, ignorance, 409,15
agnôstos, unknowable, 503,4.11.14.19.24.29.34
agôn, contest, 492,24; 493,29; 494,28, 497,16
aïdios, everlasting, 462,33; 463,1; 464,24.25; 465,24; 467,20.22.23.34; 491,21; 507,33; eternal, 518,18
aïdiotês, everlastingness, 469,1
ainigmatôdôs, enigmatically, 453,30; 454,18
ainittesthai, give riddled account, 461,11; hint at, 480,29; 499,4

aiôn, eternity, 406,11; 493,11
aisthêsis, perception, 411,7.9; 460,29; sense-perception, 439,18
aisthêtos, perceptible (thing), 453,5.8.10.15.17.24.25; 455,18; 456,13; 461,16; 469,15.24.30; 470,15; 475,23; 476,24; 477,23; 482,14.21
aithêrion, etherial, 398,11
aitia/aition, cause, 394,3.4; 395,32; 401,16; 404,23; 405,21; 406,12.13; 408,28; 419,18; 420,14; 421,26.27; 423,15 etc.; explanation, 412,33; 418,5; 432,26; 460,6; 468,8; 486,12.13.18.21; 494,22; 495,16; 496,8; 504,25.27.33; 505,24.25; 507,6.18; reason, 432,4; 465,5; 468,5; 485,22; 504,4
aitian, accuse, 437,29; 515,1; ascribe, 455,33; postulate, 450,35; 512,15
aitias, caused, 431,12
aitiôdês, causal, 460,31; causal relationship, 463,8; causal principle, 463,11
aitiologia, explanation, 408,32
aitiologikos, explicative, 496,7
akatallêlia, inconsequence, 429,27
akhronos, instantaneous, 487,15.16
akinêsia, rest, 419,9; lack of change, 435,12.13.14.15.24.25.26; immobility, 485,22
akinêtos, unchanged, 405,37; 406,2; 420,2.7.15.17.18.22; 421,31; 423,6; 435,11; 442,19; 450,5; immobile, 485,21.34; 487,12
akolouthein, follow, 397,22; 423,33; 444,28; 447,12.13.22.30; 461,30; 462,14; 463,14; 471,14; 474,12; 479,12; 484,36; 491,8.10
akolouthia, consequence, 394,14 447,17; 448,8.29
akouein, hear, 404,14; 409,25; 421,18; 439,19; read, 414,17;

490,20; interpret, 414,22; 423,14; 428,23; 446,16; 449,10; 457,25; 458,1; 474,8; 484,15; 490,19; 518,15.20
alêtheia, truth, 440,35; 468,13
alloiôsis, alteration, 397,16; 403,20; 411,5; 413,3; 415,25.31.33.36; 417,24; 419,12; 420,12; 433,1; 442,35.37; 443,12.16.17; 449,27; 450,30.34; 451,1
alloioun, alter, 423,20; 442,30.31.33.34; 510,5
alogos, non-rational, 430,19
amereia, simplicity, 514,30
amerês, without parts, 473,13; 487,26; 507,16.17
ametablêtos, without transformation, 405,29; 424,1
amorphia, shapelessness, 406,32
amorphon, formless, 480,20; 513,25
anabainein, go back, 425,2; ascend, 447,30
anaballein, postpone, 408,30.33; 420,8
anabasis, rising, 397,33; 407,18
anagein, carry up, 476,6; bring, 513,13
anagignôskein, read, 443,10
anagnôsis, reading, 427,35
anagraphein, include, 429,4; write up, 453,30
anairein, refute, 403,5; 404,18; 445,3; 464,20; 472,29; 489,8; 508,1; 510,23.25.26.30; 511,14.30; get rid of, eliminate, 483,24; 494,2
anairesis, rebuttal, refutation, 424,4; 488,2; removal, 481,25
anakampsis, bending back, 499,17
anakhusis, effusion, 503,32
anakuklôsis, recirculation, 499,17; cyclical, 500,21
anakineisthai, be activated, 426,26
analogein, make analogous, 407,20; be analogous, 446,1; 472,24
analogia, analogy, 503,5
analuein, decompose, 480,28; analyse, 507,14; resolve, 515,1
analusis, decomposition, 480,10; analysis, 507,12.15
anankaios, necessary, 394,14; 395,17; 397,5.6; 432,8; 433,4.9; 445,22; 448,9; 452,25; 465,16.25; 510,25; 515,23.25
anankê, necessary, 394,10.11.19; 438,26; 439,8.24; 442,6; 447,8.19; 448,3.6.18.24; 450,9; 451,13; 455,4; 460,8; 465,3; 468,29; 483,37; 485,34; 486,18; 490,9; 505,34; 509,3.7.14; 510,23; 511,9; 514,16; necessity, 430,22; 436,31; 465,13
anaphora, back-reference, 416,17
anaplattein, fashion, 466,1
anaploun, dissolve, 441,11
anaskeuastikos, destructive, 473,11
anaskeuê, refutation, 484,21
anatithenai, attribute, 465,13
andrias, statue, 398,11; 399,1.3.15.25; 406,31; 426,9; 492,27; 493,31; 503,8
anekleiptos, inexhaustible, 494,16.29; 497,23.25; 515,24.26.28.32.34; unceasing, 491,19; (***to anekleipton***), inexhaustibility, 396,28; 503,28
anellipes, inexhaustibility, 468,26
anesis, decreased tension, relaxation, 453,33; 455,1
anikhneuein, follow on the trail, 413,10; trace out, 456,29.35
anisos, unequal, 431,11.19; 433,15.35; 434,1.10; 444,2; 456,2
anisôsis, becoming unequal, 433,17
anisotês, inequality, 401,15; 428,26.29; 430,35; 431,5.17; 432,27; 433,16.20.25
anômalia, variation, 433,20.25.31
anomoioeidês, of different kinds, 482,34; 483,2.6.26.33
anomoiomerês, diverse, 482,7; with different parts, 483,6.7.26.27; 484,24.26; 485,7.9
anomoios, unlike, 434,7; dissimilar, 458,5; 460,15; 483,28.29
anomoiotês, dissimilarity, 401,15; 433,24
anônumos, unnameable, 430,23
anteipein, make objection, 430,4
anteirein, speak against, 432,22
anthrôpos, man, 400,32; 403,15.27; 413,22.27.29; 417,13 etc.
antidiaireisthai, be divided off, 399,20; be distinguished, 423,16; 469,11

antidiastellein, contrast, 476,28
antidiastolê, opposition, 473,31
antigegrammenos, a work written against someone, 489,22
antigraphê, copy, 399,34
antigraphon, copy, 414,19; 422,20; 427,34; 441,30; 495,8
antikeimenon, opposed, 397,32; 400,33; 424,19; 441,37; opposite, 407,12.19; 417,27; 431,25; 444,21; 484,10.11; 502,9; contrary predicate, 419,21
antikeisthai, be opposed, 400,27; 401,26; be contrary, 514,2
antikineisthai, be changed reciprocally, 419,18.19.29.31; 434,35.36; 435,1.4.18.19.34; 436,6.10.18.23; 438,12
antilegein, attack, 431,7; raise objection, 441,34
antiphasis, opposing view, 491,12
antiphatikos, contradictory, 451,27
antistrephein, convert, 424,21
antistrophê, correspondence, 402,3
antistrophon, counterpart, 468,6
antithesis, opposition, 407,9; 408,10; 428,29.31; antithesis, 412,2; 451,29; negation, 424,21
antitithenai, contrast, 407,6; 473,33.34; 474,1; 494,1; oppose, 399,30; 400,2; 401,24
anupostatos, figment, 465,33
aoratos, invisible, 470,7; 471,30.34; 472,6.18.19.20.21.22.23.24
aoristainein, be indefinite, 434,2
aoristos, indefinite, 428,26.28.30.31; 429,1; 431,13.33; 432,3.5; 433,30 etc.
aparithmein, enumerate, 499,37
apartizein, use up, 496,26; exhaust, 497,4
apatasthai, be deceived, 439,22; be misled, 516,17
apeinai, be without, 432,25; be missing, 501,12.17.18.26
apeiria, limitlessness, 451,29; 452,35; 455,2; 456,10.12; 457,1; 495,5; 500,21; 503,22.24.25.31.35; 504,7.9.16.17.27; 505,27.31; 506,26.27; 508,35.[38]; 513,17.22.29; 514,5.6.17.19; 518,4.14

apeiroeidês, unlimited in kind, 456,17
apeiron, unlimited, 394,18.19; 395,4.10.16.18.19; 396,10.23.25.26.27.29.31; 397,12; 410,10; 451,15.16.17.26.37 etc.; limitlessness, 466,3; 506,14.16.29.32; 510,4; 513,23.24.30.34; 514,20.21; 515,3
apemphasis, absurdity, 433,8
aphairein, take/off/from/away, 466,19.23; 467,7; 474,28; 479,3; 496,15.35; 497,1; 510,15.30; 512,23.26.32; 515,15; subtract, 466,29; abstract, 477,25
aphairesis, subtraction, 468,12
aphorizein, distinguish, 453,13; 484,6.9.15.16
aphorismos, delimitation, 413,5; distinction, 484,18
aphrastos, ineffable, 404,25
aphthartos, indestructible, 464,17.22.24.26.31; 465,4.14
apithanos, implausible, 450,31; unpersuasive, 463,15
aplanês, fixed (sphere), 395,11; 467,26; 482,20
apodeikunai, show, 408,7; prove, 420,8.19; 421,15; 427,28.36; 466,13; 469,30; 481,1; 512,27.33; 515,21; demonstrate, 427,25; 469,3; 511,7.21
apodeiktikos, apodictic, 476,29
apodeixis, proof, 426,22; 433,13; 476,30; 478,20.27; 480,23.35; 482,6; 511,26; 512,22; demonstration, 418,2; 440,32; 511,24; 512,26.30
apodidonai, produce, 404,12; explain, 417,4.12
apodosis, statement, 459,14; account, 422,13; 449,23; 485,35
apologeisthai, defend, 437,31
apologismos, defence, 450,31
apophainesthai, give an account, 502,6; make a statement, 517,32
apophasis, negation, 470,8; 471,29
aporein, be puzzled, 408,5; raise problem, 421,8; 468,11; 480,18; 498,28
aporia, problem, 408,31.33; 418,6.7.11; 421,9; 440,20; 441,3.16; 444,16; 445,12.14; 448,8; 467,6;

468,19.23; 480,19; investigation, 439,10; difficulty, 440,18; 445,9
apoteinein, allude, 403,2
apoteleisthai, complete, 440,33
apotelesma, effect, 419,17; result, 442,32
apotithenai, assign, 404,33
apousia, absence, 514,1
apsukhos, inanimate, 421,10
aretê, excellence, 406,33; 407,2; 409,14
arithmein, count, 477,26.32; 506,1
arithmêtikos, arithmetical, 457,15; 458,3
arithmos, number, 397,20; 448,13; 452,5; 453,6.7.9.12.16.17; 454,10.13.15.23.25.26.27.28; 455,2.3.4.6.8.9.20.26.27 etc.; (***kat' arithmos***), numerically, 450,29; 493,11
arkhaios, ancient, 399,34; 428,24; 467,4
arkhê, principle, 394,4.6.7.8.9.20; 395,23.31; 417,10.11; 420,11; 422,34; 428,28.29; 429,3.5; 440,26; 451,21; 452,5.24.25; 453,2.25.27; 454,22.27.28; 455,3.4.5.9; 456,3; 458,21; 460,6.32.36 etc.; origin, 397,16; 433,31; 438,23; beginning, 398,8.32; 400,21; 414,16; 428,18; 451,21; 460,31; 461,24.26; 462,20; 463,4.8.11.21.23.24.25.28 etc.
arkhêgonos, original, 461,18
arkhikos, being a principle, 464,29; principal, 499,37
arkhikôs, as principle, 499,32
artasthai, depend, 437,34; 448,27
artios, even, 429,11; 453,17; 454,10.11.13; 455,15.20.21.25.27.34; 456,3.4.7.10.12.14.17.20.25; 457,11.16.22.23.25; 458,2.4.5.6.11; 475,8.14.17.19; 476,17.18
asapheia, unclarity, 427,35; 429,27
asômatos, incorporeal, 418,3; 438,8.9
astheneia, weakness, 461,25
asummetria, unsuitability, 470,29
asummetron, incommensurate, 444,2
asunkhutos, unconfounded, 404,26; unconfused, 406,6
ataktos, disordered, 422,6
ateleia, incompleteness, 417,21; 435,7

atelês, unfulfilled, 398,20; imperfect, 450,15; incomplete, 398,25; 400,29; 406,34; 407,6; 414,28; 415,6.7.8.9.10.11; 416,32.34.35.36; 417,16; 426,19; 428,9; 429,1; 430,33; 431,15; 432,14; 433,22; 434,27.28; 437,15
ateleutên, incompletely, 470,6.17
ateleutêtôs, without end, 454,3
athroôs, all at once, 437,12.14; 494,17; 495,2; 497,15; instantaneously, 463,21; 464,32; 494,34
atmêtos, uncut, 454,1.2; 511,8.23; undivided, 495,24
atomos, individual, 405,20.21; 490,5; 494,9; particular, 490,8; atom, 453,3; 459,18.27.28; 460,5; 461,34; 462,5.12; 468,30; indivisible, 492,2.4.7.8.10.11
atopia, absurdity, 443,31; 444,30; 503,21
atopos, absurd, 410,17.18; 417,19; 424,1.22.27; 425,12.13; 427,15; 431,20.21; 437,16; 441,22.24.28.38; 442,17.36; 443,15.20.21; 444,14.17.23; 445,7.21; 446,23.25; 467,28; 471,15; 474,26; 483,23.31; 487,11; 488,19; 491,29; 504,16; 517,1; 518,18; strange, 438,4; 465,15
aülos, immaterial, 398,7; 400,15; 503,25; 514,30
autarkês, sufficient, 397,35; 431,1; independent, 485,34
autokinêtos, self-changing, 402,16; 403,3; 421,5.15.25.27.28.33; 423,11; 440,27
automaton, automatic, 462,29.31.34
autotelês, self-sufficient, 406,7
auxanein, increase, 396,33; 442,33; 450,25; 456,20; 460,17; 467,9.11.28 etc.
auxêsis, increase, 396,24.33; 407,15; 468,16.32; 475,21; 494,15; 499,15.21.25; 502,32; 505,28.33; 506,2; 509,11; 510,20; 511,19; growth, 397,32; 403,21; 413,4; 416,3; 417,26; 422,28; 431,32; 491,21
axiologos, worthy of mention, 452,23
axiôma, axiom, 397,36; 466,10;

Indexes

482,13.21; 485,17; assumption, 406,19

baros, weight, 486,13; 488,10.12.16; heaviness, 489,6
barunein, weigh down, 407,1; give a barytone accent, 443,11; 490,17
barutonein, make definite, 490,29
batheôs, fundamental, 421,8
biazesthai, be forced, 423,2
biblion, book, 394,3; 395,15.21.27; 397,23; 398,32; 399,7; 400,5; 413,8; 417,8; 450,33; 463,7; 479,19; 480,35; 485,10; 487,15; 492,3; 507,16; 512,34
bôlos, clod, 422,35; 482,24.25; 483,10.11.20; 486,27
brakhulogia, succinctness, 495,15

dapanan, expend, 496,1
deiknunai, show, 397,12; 401,8.34; 402,4.18; 405,6; 407,21.30; 416,10; 421,30; 423,28; 432,3; 437,23; 438,12.16; 445,21; 451,25; 542,23 etc.; demonstrate, 397,26; 398,33; 401,23; 402,32; 408,19; 426,13.20.30; 437,14; 449,32; 482,3; 484,28.35; 492,6; 512,36
deiknumenon, demonstration, 510,37
deiktikon, proof, 472,36
deixis, proof, 475,4; 489,1; 491,25; 492,4; 510,24; 511,10.25; demonstration, 531,14
dekhesthai, receive, 398,11; 399,14.15.16.18; 400,30; 467,19.25; 506,4.29.32; 508,34; 513,17; 514,1
dêloun, show, 396,8; 398,15; 399,36; 401,30; 419,16; 428,24; 432,4; 446,18; 457,7; 470,25; 476,30; 483,32; 486,3; 490,19; make clear, 404,1; 405,29; 451,21; 454,6; 474,9; 482,15; 485,12; 489,16; 490,27; 497,4.27; 502,6
dêmiurgikos, demiurgic, 461,14
dêmiurgos, creator, 507,32
Dêmokriteios, of Democritus, 459,27
diadokhê, succession, 494,31
diagramma, diagram, 511,16; 512,27
diagraphein, strike out, 428,2.3
diairein, distinguish, 395,26; 399,28.32; 408,2; 413,12; 417,9; 469,32; make distinction, 408,18;

417,7; 463,7; divide, 397,30; 399,23; 400,9.11.14.20.23; 401,2.9; 405,11; 439,12; 449,26; 451,28.30; 454,5; 455,5.21.22.28.29.31; 456,1
diairêma, divisum, 455,32; 456,11; section, 492,16; portion/segment, 494,32; 495,25.34; 504,10; 505,32; 506,11.28.34; 508,26
diairesis, being divisible, 396,27; division, 398,1.16; 399,37; 400,13.21.24; 401,8; 427,12; 446,11.21.25.29.36; 442,24; 444,7; 451,27 etc.
diaitêsôn, arbitrator, 491,9
diakosmêsis, imposition of order, 461,11
diakosmos, universe, 404,23
diakrinein, distinguish, 404,23; 413,20; 435,17; 457,14; 460,25; 461,2.19; separate, 432,31; 436,11; 460,22; make a distinction, 410,32; 461,5.21
diakrisis, distinctiveness, 406,7; distinction 460,31; 461,2.3.21.23; 503,26
dialektikê, dialectic, 476,26
dialusis, solution, 408,33; resolution, 480,8
diametron, opposition, 419,34; diagonal, 457,4.5
dianoên, object of thought, 475,24
dianoia, mind, 512,22
diaphanês, transparent, 425,19.20.21
diapherein, differ, 395,26; 405,12; 410,35; 444,23; 450,29; 453,4; 465,24; 467,29; 483,35; 498,6; 511,11
diaphônia, disagreement, 404,21
diaphora, difference, 405,4; 411,29; 422,26; 428,10; 430,14.15.16.18.20
diaphoros, different, 395,21; 403,36; 408,3.26; 409,31; 412,26; 419,12.16; 425,17; 426,10; 433,31.37; 437,6; 442,29.32; 443,33; 446,4.27.31.32; 447,2.6.23; 448,18; 449,27; 452,26; 470,18
diaplasis, moulding, 445,29
diaptuxis, unfolding, 438,3
diarrhipsis, fragmentation, 487,29
diarthrôsis, analysis, 439,4
diaseiesthai, be disordered, 422,6
diastasis, conjunction, 419,16;

interval, 447,28; extension, 448,12.14.17.25; 449,16; 479,16; 507,17
diastêma, interval, 439,33; 446,31; 482,17.18; 512,24
diataxis, classification, 412,33
diathesis, state, 400,38; condition, 411,14; 445,32; disposition, 419,13
diatithenai, give condition, 426,29; have character, 440,7; be subject, 441,32; give disposition, 450,17
didaskein, teach, 395,9.15; 405,27; 442,9.31.32.33.34.35; 445,1.24.26.34.36; 446,2.3.28; 447,8.13.14.15.16.19.20; 448,1.3.4.6.7.10.13.15.20.24; 449,14; 451,18; 464,26
didaskôn, teacher, 446,12.18; 447,5.12.14
didaxis, teaching, 442,26; 444,18.28.33; 445,1.24.30; 446,26.28; 447,12; 448,9.12.19.22.23.28.31; 449,6.7.9.12
didonai, give, 396,34; 466,11; 484,20; 492,9; 498,25; 508,18; 510,22; 511,8.23.32; 512,1.18
dielenkhein, refute, 401,13; 448,8.28; 487,15; 515,20.22; criticize, 428,24; examine, 478,25
diexerkhesthai, traverse, 476,13; 477,27
diexitêtos, can be traversed, 470,28; 477,16; traversable, 470,31; 477,36
diistasthai, extend, 448,13.15.18.25; 478,6; 479,9.11; 502,7; 514,12; separate, 487,24.28
dikaiosunê, justice, 476,4.6
dikhotomia, bisection, dichotomy, 455,22; 508,22.23.26.29.31.36; 509,3.9.10
diorismos, distinction, definition, 444,16; 450,3; 501,32
diorizein, define, 429,25; 402,3; distinguish, 435,9; 436,22; 469,27; 476,32; 489,15; 516,21
diorthoun, correct, 500,8
dogma, opinion, 404,17
doxa, opinion, 420,9.13; 424,36; 428,21; 430,29; 431,7; 445,10.19; 452,22.27; 460,6; 461,11; 463,15; 465,26; 480,29; 485,28; 514,33

drastêrion, activity, 411,8
duadikos, dyadic, 454,34
duas, dyad, 453,25; 454,8.10.13.14.15.28.29.34.35; 455,2.5.8.9; 456,3; 458,14; 499,4.31; 503,7.26
dunamis, potentiality, 398,25; 400,33.34.36; 409,29.30.31; 416,32.36; 424,8.10.12.17.28.29; 432,8.12; 433,32 etc.; power, 405,28; 453,22; 462,18; 467,7; 479,1.5.6; 481,27; 482,9; 486,10; 488,6; 489,5
dunamei, potentially, 398,2.6.9.10.18.19.21.26.31.33.35; 399,1.2.3.7.11.12.16.17.20.21.23.24. 27.29.31.33.35.37 etc.
dunatos, possible, 404,34; 442,31.33; 443,10.13; 458,1; 468,34; 470,2; 472,25; 474,25; 475,34; 476,13; 480,26; 483,26; 490,20; 498,5.16.34; 499,2.26.28; 505,35; 507,21; 508,4.7.15.18.29; 510,22; 511,8.10.11.13.26.31; 513,28; 516,8

eidenai, understand, 394,6.20.23; 400,24; 428,18; know, 394,10; 395,20; 422,24; 423,20; 428,3.5; 431,6; 436,13; 472,25; recognize, 421,13; 499,35; 501,16
eidêtikos, formal, 429,7; 513,34
eidikos, specific, 395,27; 402,20; formal, 513,11; 514,29
eidopoios, supplying form, 456,16
eidopoioun, inform, 454,16; give form, 479,22.24; 481,22; 503,23
eidos, species, 395,34; 397,29.36; 401,9.31; 403,7.24; 404,3; 405,10.19.20; 408,1.6.9.11.26; 409,11; 413,6 etc.; kind, 478,16; 482,25; 483,21.35; 484,26.29; 485,9; 489,12; 501,36; type, 401,34; 417,23; form, 398,8.9; 399,9.10.15.17.18.26; 400,15.18.33; 406,18.21.22.28.31.32.33.35; 407,5.9.10.19.23.32.33.34 etc.
eikôn, image, 394,24
eikotôs, reasonably, 394,4; 397,7; 400,19; 404,24.27.37 etc.
eisagein, introduce, 458,27.31; 459,13.18; 460,2; 462,25; 468,30;

Indexes

512,12; 516,5; 518,3; make statement, 510,17
ekballein, produce, 510,21; extend, 511,1.25.32; 512,9
ekdokhê, sense, 428,13; interpretation, 503,11
ekei, THERE, 406,6; 503,16.19.25.27.31.32
ekkrinein, distinguish, 460,23; separate out, 460,35; 461,7
ekkrisis, being separated out, 460,20; separating out, 461,32
ekpurôsis, conflagration, 480,29
ekstasis, turning from, 413,30; departure, 451,6
ekteinein, stretch out, 467,27.28.29; 468,1.3; stretch over, 487,26
ekthesis, exposition, 495,14
ektithenai, offer, 417,23; set out, 433,13; 452,27; 456,22.24; 470,16
elakhistos, smallest, 494,4.7; least, 496,1; 508,10; minimum, 505,23; 509,19
elenkhein, examine, 437,15; 478,19; 503,11; refute, 468,32; 479,19; 486,21.22; 515,23
elleipsis, deficiency, 401,6.10.13.14.30.35; 443,36; 454,12.13.16.32; 553,27
emperiekhesthai, contain, 454,11
emphainesthai, appear, 396,8.10.22; 439,10; make its appearance, 404,5.7
empsukhia, animation, 421,12
empsukhos, animate, 405,37; 421,7
enagein, include, 460,20; set out, 472,26
ta enagonta, considerations, 466,28
enallattein, alter, 456,21
enantios, opposite, 406,26.27.30; 407,6; 408,35; 409,2.15; 410,5.6; 411,24.30; 413,4; 416,4.7; 424,8; 425,6.10.13 etc.; contrary, 420,8; 504,31; 513,22
enantiôs (ekhein), relation of opposition, 480,3
enantiôsis, opposition, 409,10.13; 410,28; 411,35; 412,1.4.5; 443,34; 445,2; 479,24; 481,12.28.33; 484,4.6; opposed, 435,36
enantiotês, opposition, 409,14; 410,4
enargês, obvious, 396,7; 416,11; 423,33; 441,5; 444,30; 446,33; 447,14; 451,8; 474,25; 477,19; 483,27; 488,12.19; 429,27.29
endeiknunai, make clear, 415,29; 497,19; show, 468,27; indicate, 444,12; 457,15; 459,19; 488,33; 500,30; 506,18
endeixis, pointing, 442,2
endekhesthai, can be, 399,4; 467,21; 476,9; 477,26.27; 509,16; 517,10; be possible, 409,2; 442,4; 461,20; 472,34; 473,9; 474,2; 475,6; 480,21; 482,6; 507,27; 509,19.20; 515,27; be capable, 426,30.32; 428,15; 429,31; 507,26
endidonai, allow, 443,35; refer, 450,4; communicate, 450,10
endoxon, accepted opinion, 440,22; 476,27
energeia, activity, 398,6; 401,18.19; 404,34; 405,2.28; 406,10; 407,25; 408,16; 414,20.25.27.28.33.37; 414,4.6.7.8.9.10.11.12.18.19; 418,14.15; 419,3; 420,25; 421,17.24; 422,4.8.9; 424,2; 426,30 etc.; actualization, 408,20.22.30; 413,13; 414,5.12.23.32; 415,16; 416,12.14.16.21.23.35; 417,31; 418,10.16; 422,15.17.32; 423,33
energeiai, actually, 396,27.30; 400,17.32; 413,26.28.29; 414,1; 417,34; 418,1.2.10.11.26.31; 419,1; 422,17; 424,1.8.11.19.22; in actuality, 398,10.19.21.31; 399,8.13.14.16.17.18; 416,20.25; 419,23; 422,14.16; 508,13 etc.
energein, be actual, 400,31; 413,32; 440,6; 494,10; exhibit (actualization), 416,35; be active, 400,33; 414,11.14; 415,5.17; 418,21.27.29.32.33; 419,2; 422,10.21.23.30; 424,2.3; 426,25.33; 427,5; 428,6; 436,16; 439,25; 447,3; 450,1; act, 405,3; 422,31; 427,1; 435,30.33; 437,25; be actualized, 416,19.24.25; 418,22.23; 426,24.31; 427,3; 436,5; 438,1; actualize, 442,27; 446,22; 447,2
energêtikos, the active, 401,29; 426,28; activates, 439,28; of activity, 420,33
engraphein, inscribe, 466,13

enistasthai, attack, 445,10; reject, 515,24
enkalein, accuse, 429,27; 433,7
enkhronos, temporal, 461,12
ennoein, bear in mind, 411,10; 443,27; conceive, 473,34
ennoia, concept, 406,15; 465,26; 470,22; 477,25; 513,29; thought, 477,9; 490,34; idea, 428,24; 460,11; 461,5; 496,5.12; 499,2.3.10.18; 500,8; conception, 432,4; 512,12
enokhlein, trouble, 426,12; bother, 440,21; embarrass, 510,36
enstasis, objection, 422,12; 426,12; 431,23; 445,16; 447,21; 478,35; 489,30; 492,19
entautha, HERE, 404,28; 412,11; 503,29; our world, 406,9
entelekheia, actuality, 397,25; 398,2.3.4.6.7.8.9.12.17.18; 399,20.21.22.25.26.29.30.31.33.34. 35.37; 400,1.4.6.9.12.13.14 etc.; actualization, 401,23; 409,26; 412,21; 413,7.9; 414,18.19.20.22.29.30.31.36; 415,2.3.5.7 etc.
enulos, material, 404,27; enmattered, 480,14.15; 503,14.21; 505,30; 506,17.34
enuparkhein, be within, 419,13; 460,10; be present in, 460,24; 461,33; 480,17; 504,9; 506,15.16.34; 513,27; be immanent/ inherent, 491,2; 497,19; 508,30
epagein, add, 405,1; 407,31; 408,1.11; 420,6; 427,30; 429,28; 432,1; 436,15; 442,4; 443,21; 444,13; 450,34; 458,11.35; 469,22; 474,8; 505,24.29; 511,25; 515,34
epagôgê, induction, 513,12
epamphoterizein, be ambivalent, 484,12
epiblepsis, perspective, 482,10
epibolê, viewpoint, 406,13
epidekhesthai, admit, 410,3
epideiknunai, demonstrate, 404,21; exhibit, 499,1
epidiamenein, survive, 441,2
epidosis, increase, 416,3
epigraphein, entitle, 398,33
epikheirêma, argument, 472,27; 473,11; 476,32; 491,5

epikheirêsis, argument, 440,26; 472,30; 473,1.17; 482,9; 488,6; 495,14; 516,21
epileipein, lack, 415,31; end, 494,31; 496,13; 515,14.16
epinoein, conceive of, 403,1; 467,15; 511,27.29
epinoia, concept(ion), 467,12; 468,8.15; 506,5
epiphaneia, surface, 454,25; 487,23
epipherein, add, 429,30; 443,18.22; 488,19
epiphora, supplement, 515,34
epipolaios, manifest, 442,7; superficial, 503,11
epirrhêma, addition, 446,20
episêmainein, exhibit, 415,31; indicate, 438,17
episkepsis, overview, 395,8
episkopein, survey, 475,26
epistêmê, knowledge, 401,33; 409,15; 476,4.7; science, 451,10.22; 452,12.19.21
epistêmôn, wise, 469,24; man of understanding, 475,31; 510,17
epistêtos, known, 401,33
epistasis, increased tension, 453,33; increase, 455,1
epitêdeios, suitable, 417,1
epitêdeiotês, suitability, 400,29; 417,1; 439,33; 465,21; suitable condition, 400,37; readiness, 476,19.26
epizêtein, seek for, 493,21.22
epizeuxis, joining, 511,2.5
eran (v.), love, 440,35
êremein, be static, 415,3
êremia, rest, 411,21; 432,11; 435,13.14.16.24.26; 483,23
ergon, action, 400,35; product, 491,8.9
erôtan, raise question, 467,26; ask question, 467,31
erôtêsis, question, 418,8; 424,13
eskhaton, extremity, 467,27; the outside, 488,21.33.34.35.36
ethos, custom, 457,16
eugnômonês, wise, 422,23
eulogôs, appropriately, 396,21; reasonably, 437,20; 449,33; 462,23.27; 464,19
euoristos, easily distinguished, 481,32
euruthmia, good form, 442,20

euthetismos, ordering, 487,29
euthunein, examine, 431,18; 476,18; 487,33; contest, 485,28
eutheia, (straight) line, 492,9; 500,20; 510,21.35.36; 511,1.3.8.11.23.31.32
eutupôn, easily takes an impression, 481,24
exaireisthai, transcend, 435,8; remove, 495,12
exêrêmenon, transcendent, 405,21; 491,1
exanastasis, separation, 405,28
exêgeisthai, explain, expound, interpret, 409,25; 414,17; 428,4.5; 440,15; 446,17; 449,5; 459,5; 469,10; 472,8.25; 501,2; 517,17.20; give an account, 436,19
exêgêsis, explanation, 457,13; interpretation, 472,27; 485,12
exêgêtês, commentator, 395,14; 406,28; 408,30; 432,21; 471,35; 472,8; expositor, 455,21.25
exetazein, examine, 512,25
exisazein, be equal (extension), 437,28; 478,30; 489,18

genesis, origin, 396,23; 432,27; beginning, 515,27; generation, 422,7; 515,24.26.35; coming to be, 397,16.30; 403,19; 407,14.34; 408,4; 409,11; 410,14; 412,9; 413,4; 416,6; 417,6.8; 419,8; 435,2.17; 436,18; 460,20.34.35; 461,34; 463,33.35; 464,5.32.34; 465,7.9; 466,29; 480,12; 481,16.17; 482,11; 494,16.22.26; 497,23; 505,36; 515,14.29.32
gennan, procreate, 427,2
genos, genus, 396,13; 400,17; 402,21.24.25.27; 403,4.6.7.10.12.13.19 etc.; kind, 399,28.32; 400,12.20.24; 404,37; 407,12.30.31; 408,14.16.17; 413,3.12.20; 432,16.17.18.28; 433,18; 435,35; 509,27.28; 516,11
geômetrês, geometer, 510,19; 511,18
geômetria, geometry, 437,3; 468,30; 510,18.25.27.29; 512,30.34
gignesthai, take place, 395,1; exist, 468,33; result, 457,19; 466,19; 508,29; 509,3; come to be, 398,8; 409,16; 410,12.15; 411,26.27; 412,17.19; 414,7; 431,15.16; 454,2; 458,22; 460,16.20.34; 463,34; 464,27.28.31.32.33; 465,1; 466,30; 467,13; 480,15.32; 481,8.10; 487,26; 492,16.17; 493,16; 494,30; 495,2; 497,34; 498,4; 505,35; *ta gignomena*, things, 420,12; 459,4; 514,36; results, 448,33; 449,4
gignôskein, have knowledge, 394,7; understand, 394,9.11.16; recognize, 412,32
gnômê, view, 413,10
gnômôn, gnomon, 457,1.3.4.6.9.10.12.14; 458,1.6.7.8
gnôrimos, plain, 449,29; 450,30; familiar, 460,29
gnôsis, knowledge, 394,8; 432,15; 445,33; 503,2
grammê, line, 452,4; 454,23.24; 462,8; 466,11; 492,2.4.6.8.10; 510,24.33.35; 511,25; 512,5.13.19.27; 516,26.27.28
grammikos, graphic, 512,32
graphê, reading, 395,21; 423,13.21; 436,19; 446,19; 495,9; text, 399,34; 414,19; written, 441,31
graphikos, scribal, 481,33
graphein, write, 400,5; 413,5; 431,3; 432,21; 436,13; 437,7; 440,13; 441,30; 451,22; 453,31; 454,19.21; 492,3; 496,12; draw, 511,32; 512,27; paint, 517,11.12

hamartanein, err, 425,34; 426,2
haphê, touch, 438,9; contact, 458,18.28; 459,20.22.24; 460,3; 462,2.16
haplôs, simple, 397,17; 399,26; 417,2.18
haplôs, simply, 398,7; 399,2; 410,24.26; 416,6; 423,3; 427,9; 432,10; 434,26; 445,4.5; 446,14; 459,8; 479,8; 490,8.9; 499,22; 513,29; 517,23; unqualified, 501,3
haptesthai, touch, 438,7.8.11; 459,24; 462,15; 483,30.31; 516,1.2.9.10.12.17.18.22.23.24.25.27.28; study, 510,29
harmottein, be suited, 440,34; apply, 476,21
hêgeisthai, think, 431,33; 454,27; believe, 460,11

hêgoumenon, antecedent, 424,4; protasis, 245,13; 489,32
(to) hen, the one, 420,13.17; 429,12; 453,25; 454,14.15.28.34; 455,3.5.7; 481,13.15.17.22; 483,30; 484,15; 503,30; 505,8; unit, 458,8; 503,17; 504,34.35; 505,2.4.12.14.15
henas, unit, 505,10.11; 507,20
henoeidês, uniform, 501,34
henôsis, unity, 404,25; 406,7; unification, 458,28; 459,21; 461,19; 503,32; union, 475,24
henousthai, be unified, 461,12; 487,20; 501,34
hepomenon, consequent, 424,4.6; 485,8; consequence, 443,9; 444,14.17; 487,17; 488,19.26; conclusion, 489,32
hermêneia, interpretation, 400,4
hermêneuein, interpret, 428,18
hetairos, associate, 411,16
heteroiôsis, becoming other, 430,7; 432,37; 434,6
heterokinêtos, externally changed, 421,4.25.26; 460,36
heterotês, otherness, 401,15; 404,24; 428,25.29; 429,31; 430,1.4.6.8.10.24.27.35; 431,17.27; 432,34.36.37; 433,2.5.7.11.25; difference, 424,19; 441,6.19
hêtusthai, be defeated, 467,8; 478,34
heuriskein, discover, 405,30; 406,4; find, 414,19; 464,4; 466,23; 485,21
hexagônikos, hexagonal, 419,14
hexis, disposition, 414,32.33.34; 446,1; 461,22; condition, 432,8; 450,9.11.16.26; possession, 451,28
histanai, remain (constant), 406,1; 413,33; 430,14; 434,3; halt, 455,28; 471,4; 505,22; 507,20; set, 500,1
hodeuein, progress, 434,18
hodos, journey, 407,19; road, 416,31; 447,26; route, 426,25, 470,27
holos, whole, 421,13; 425,4; 447,29; 450,11.14; 482,24; 483,11.20.36 etc.; **(to holon)**, the whole, 417,12; 457,6; 460,30; 465,3; 466,21; 474,23; 482,23; 483,36; 484,1.16.24.30; 486,24.26.34; 487,15 etc.; universe, 460,8.9; 485,33; 489,15
holôs, in general, 399,2; 401,17; 402,1; 415,9; 417,13; 427,15; 437,34; 449,14; 450,1; 453,7; 455,38; 462,22; 470,5; 476,8; 487,31; 493,34; 497,14; 499,32; 505,12.35; 506,7; 512,30.31; 518,17
holoskherês, general, 397,35; 428,23
homalos, uniform, 432,25; stable, 434,10
homalotês, uniformity, 432,22.26
homoeidês, of the same form, 459,23.25; of the same kind, 462,14.15; 482,34; 483,1.6; 486,22.23.27
homogenês, of the same sort, 396,12; 440,31; generically the same, 431,30
homologein, agree, 400,10; 405,8; 409,14; 420,18; 421,7; 437,9; admit, 409,11; 411,9
homoiomereia, homoiomeries, 453,3; 459,14.17; 460,4.9; 461,32; 473,21
homoiomerês, uniform, 482,7; has similar parts, 482,28.30; 484,23; with similar parts, 482,23.36; 483,6.7.9.11.25; 484,24; 485,7
homoiotês, similarity, 401,32; 500,23; 502,19.26; 516,32; likeness, 460,30
homônumia, equivocation, 406,14; ambiguity, 463,6
homônumos, equivocal, 400,26; 404,10.12; similarly named term, 510,2; has more than one sense, 402,20; equivocally (named), 403,12.13.15.16.18.28.33.35; 404,8.9.13.15; 415,29
horan, see, 401,3; 423,5.17; 425,20.22.30.31.32; 428,14; 429,31; 432,14; 439,19; 444,30; 446,8; 454,8; 458,4; 460,12.21; 463,23; 470,7; 472,14; 480,15
horatos, visible, 398,26; 425,17.18.22.26.27.31.33.34; seen, 401,33
horismos, definition, 394,7.13.17; 396,11.18; 397,27; 401,3; 403,2; 404,9.13.14; 408,25.27; 413.16; 414,17.21; 415,29; 416,9.10; 418,7; 422,14; 425,35; 428,1; 436,25; 438,3; 439,3.15; 445,15; 447,24; 448,1; 449,24.31; 450,2; 501,15; 502,17
horiston, defined term, 394,16
horizein, separate off, 516,25; define,

394,4.18; 396,17; 402,17; 404,8;
413,25; 414,15.26; 425,21.24.25.35;
428,25; 437,33; 446,11; 449,25;
500,25; limit, 477,19; 479,26;
502,14; be determinate, 470,28;
definite, 428,32; 431,15.31; 432,18;
433,30; 446,14; 455,6; 457,22;
490,18.20.21.30; 493,16; 496,2.9.20;
505,8; 506,19.23; 507,28;
509,1.5.6.13.14.15.16; 510,14
hormê (kath'), voluntary, 422,29;
motivated, 423,4
horos, determination, 396,12;
definition, 404,9.10; 437,15.28.29;
450,27.32; 477,18; 481,31; 502,31
hôsautôs ekhein, be uniform,
420,26.28.29
hudôr, water, 399,1.13; 423,19;
450,18.19; 452,32; 453,21;
458,25.34; 459,1.3; 460,14.18.22;
471,3; 474,22; 481,30; 484,11;
514,15.34
hugiainein, be healthy,
424,9.10.17.26.31; *to hugiainein*,
health, 424,32
hugieia, health, 423,23
hugrotês, moisture, 423,24;
424,31.33; moist, 479,25
hulê, matter, 398,9;
399,7.10.13.17.24; 418,3.31; 419,22;
434,15; 435,35; 456,13; 458,14.21;
459,6; 463,9.25.26.29.30.31; 467,37
etc.; material, 419,21
hulikos, material, 456,17; 458,20;
459,11.15; 465,8.11.20;
503,22.24.29; 506,34; 513,10;
514,24.30.31.34.36
humnein, hymn, 453,11
hupantan, deal with, 442,12; conflict
with, 445,14
huparkhein, be present, 395,6;
405,15; 406,17.21; 407,22.28;
408,1.9; 443,4; 452,15; 454,9;
475,1.9; 513,30; accept, 489,19;
come from, 488,8; be, exist, 395,7;
396,6.13; 401,7; 434,17; 435,31;
445,18; 453,13; 472,21; 477,31;
479,1.5; 495,5; 498,6; 501,34;
502,28; 505,1; occur, 396,10;
403,30; 409,14; 434,24;
435,14.24.25; 472,29; 475,28.29;
509,25.26; *ta huparkhonta*,
constituents, 448,30
huparkhon, reality, 471,20.21
huparxis, validity, 490,4; being,
490,11; *en huparxei on*, existent,
490,5
hupenantios, opposite, 412,26
huperballein, exceed, 479,5; 498,12;
504,5.14.26.32; 506,19.22.28;
508,15.18.24.27.33.37; 509,3;
510,22; 511,20; 517,23; be an
excess, 484,36; 485,2
huperbolê, excess, 443,35; 509,1.5.16;
enormous, 470,21
huperdramein, spill over, 503,33
huperokhê, excess,
401,5.10.12.14.30.35;
454,12.16.31.32; 503,27; 517,7;
superiority, 461,13; 479,1
huperpiptein, fall outside, 504,11
huphistanai, exist, 397,12; 412,10;
434,25; 453,8; 473,34.35; 490,33.34;
492,23.26.28; 495,2; 499,34; 511,28;
be existent, 506,4; be, 397,13;
494,24; *kath' hauto
huphistanai*, be self-subsistent,
471,32; 473,26; 478,3
hupoballein, subsume, 452,2
hupodokhê, receptacle, 487,24
hupographê, sublime, 449,21
hupographein, outline, 397,22;
sketch, 450,28
hupokeimenon, substrate, 398,24;
410,21.23.30; 411,36;
412,1.11.13.14.18; 423,29.31;
424,9.28.30; 425,4.6.8.11.13.16.17;
426,10.13; 435,32; 437,5 etc.;
subject-matter, 409,31
hupokeisthai, be a substrate, 514,10;
subsist, 473,17; be present, 416,25
hupokruptein, hide, 456,8
hupolambanein, suppose, 395,1;
think, 395,13; 453,16
hupoleipein, leave, 446,4.30; 453,22;
exhaust, 467,8; 517,5; come an end,
493,16; *oukh' hupoleipein*,
inexhaustible, 466,16
hupomenein, undergo, 412,10;
continue, 412,19; be permanent,
518,15; persist,
506,3.5.8.10.25.28.30.33; 507,12;
513,19.20.23

hupomimnêskein, recall, 402,25; 480,35; 491,15; remind, 424,14; 449,20; 501,16; 505,4.15; 510,5.8.10; 511,6; 513,10
hupomnêma, commentary, 394,2
hupomnêmatismos, comment, 430,3
hupomnêmatizein, comment on, 430,2
huponoein, get the idea, 460,24; suspect, 472,14
huponoia, idea, 460,7; supposition, 484,8; understanding, 399,32
hupopiptein, fall under, 408,24
hupostasis, existence, 398,19.23; 406,12; 490,32; 494,23; 508,22; occurrence, 494,26; object, 512,21
hupothesis, hypothesis, 420,15; 442,13.15.35; 443,8; 445,21; 461,6; 469,25; 476,18; supposition, 472,31; 479,14.19; 484,17.19; 485,6.8.10
hupotithenai, suppose, 420,10.22; 442,25; 443,5.7; 444,14; 452,28.31; 458,17.22.24; 459,28; 460,6.18; 463,19; 465,36; 466,31; 474,13.29; 475,5.13.14; 478,21; 479,7.12.18; 483,27; 484,23; 486,7.12.20; 510,24.34; 512,35; 514,33; posit, 461,26; 465,10; 481,25
husterogenês, posterior, 491,1
husteros, subordinate, 487,10

iatreusis, curing, 417,24; 450,34
iatros, physician, 424,35
idea, idea, 402,13; 403,3.6; 455,8.9.19; 458,15; 476,1.2.3; 499,12; 503,18; form, 420,18.21.22; 413,23.25
idios, special, 449,18; own, 485,31; character, 481,31; proper, 425,24.35; 440,25; peculiar, 465,5
idiotês, special nature, 404,35; particular variety, 449,27; character, 405,23; individuality, 430,25
idiotropos, of a special sort, 433,19
ienai, move, 395,22
isakis ison, symmetry, 456,19.35
iskhuein, be capable of, 467,7; 494,2
isopleuros, equilateral, 511,33.34; 512,2
isos, equal, 403,15; 434,1.4.10; 444,2; 455,20.21.24.25.27.30.33; 456,2; 466,14.19.23; 478,35; 488,23; 502.7

isoskelês, isosceles, 466,17.21
isosthenês, equal in strength, 478,30; of equal strength, 479,3
isotês, equality, 401,32; 431,26

kairos, occasion, 408,33
kakia, badness, 406,33; 407,3; 409,14
kallos, beauty, 442,20
kamnein, be diseased, 424,8.18.26
kamptein, bend down, 407,1
kanôn, principle, 444,3
katabainein, descend, 447,31
katabasis, falling, 397,33; descent, 407,18; way down, 447,27
katadeiknunai, demonstrate, 440,36
katagignesthai, be engaged in, 458,20
katagraphê, drawing, 457,23
katagraphein, construct, 511,15
katakolouthein, follow, 421,23
katalambanein, occupy, 467,37; include, 477,32
katalêgein, be exhausted, 492,33; finish, 504,29
kataleipein, leave, 455,29
katallêlos, grammatical, 400,3
katamanthanein, understand, 433,25
katametrein, measure, 496,2
katantan, reach, 455,28; 508,12; 509,11.12; meet, 513,21; 516,6
kataskeuazein, construct, 470,30; confirm, 490,1.3
kataskeuê, construction, 470,24.32
katastasis, constitution, 435,16
katêgorein, predicate, 415,27; 474,5; 490,5; 509,27.35; categorise, 430,11.17
katêgorêma, category, 402,25
katêgoria, category, 397,26; 398,4.6; 399,23; 400,9.14.22.23; 403,9.21; 404,1.31; 408,7.26.29; 409,8.9; 410,2.33 etc.; predication, 510,2.4
katêgorikos, categorical, 425,5
katharotês, purity, 404,26
kath' hauto, self-subsistent, 473,7.8; self-existent, 425,33
kathêgemôn, master, 451,4; 517,16
kath' hekasta, particulars, 460,30; 501,24.25; particular cases, 501,16.18
kathêsthai, be seated, 408,23

katholikos, general, 440,32; 476,20
katholou, universally, 395,6; 416,22; 451,1; in general, 449,23
kathomilein, accept, 459,2
keas, cave, 470,21.24
kenembatein, step on emptiness, 465,31; be of the unreal, 508,19
kenkhris, 470,13
kenon, void, 395,1.2.3.5.12.13.16; 397,1.3.4.5; 462,1.16.17.19; 467,16.17.19.24; 484,36.37
kentron, centre, 511,35
kephalê, head, 505,6
kêros, wax, 445,28.32
khalepotês, difficulty, 432,5
khalkos, bronze, 398,10; 399,1.3.14; 406,32; 424,2.6.16; 426,9; 492,27; 503,7; brazen, 503,8
kharaktêrizein, characterize, 515,5
kheiragôgia, clarification, 416,12
khôra, space, 467,37; field, 507,7
khôrêgia, supply, 465,8; means, 540,9
khôrion, place, 395,20; area, 440,12; passage, 472,8
khôristos, separate, 434,23; 453,6; 471,14.31; 505,29; 506,20.23; 508,30
khôrizein, separate/off, 434,36; 435,1; 446,18; 462,22; 477,24; 491,1
khreia, need, 408,1; 418,5; 491,9; 494,6; advantage, 437,22; purpose, 465,11
khrêsis, ordinary use, 501,15
khrôma, colour, 397,31; 398,25; 406,24; 425,16.18.19.20 etc.; 442,20; 470,8; 512,28.31
khronikos, temporal, 411,14; 460,32; 461,23; 464,21
khronos, time, 394,22.24.25; 395,5.11.17; 397,1.9.11.18.19.20.21; 406,10; 410,36; 411,4.5.6.8.9.10.11.18.19.20.21 etc.
khumos, flavour, 406,24; juice, 424,34
kinein, change, 396,6; 402,12.17; 418,6.9.29; 419,17.18.19; 421,16; 438,27.33; 440,4; initiate change, 418,12.17.23.24; 420,1.2.6.7.21; 421,25.31; 422,1.2.18; 423,3.5; 426,13.28; 432,23; 434,34.35; 435,1.2.3.6.10; 436,8.9.10.16.18.22; 438,18; 439,25; 440,11; 450,4.5.8.11.14.16.22.23.24
kinoun, changer, 401,15.18; 419,27; 420,21; 421,30; 423,12.15; 433,21.24.27.31; 434,32; 436,6.14.21; 437,7.8.11.12.14.17.21.27.29; 439,2.6.9.11.12.21.24.25.26.29.35.36; 440,3.16.20; 441,4.12.21.23.27; 442,14.16.18.19.30.37; 443,1.27; 444,6.19; 445,23.27; 446,1.10; 447,11; 448,36; 449,32.36; 450,5.9
kineisthai, change, 395,3; 397,10.17.18.28; 401,16; 402,10.12.16.18; 404,27; 406,1; 409,6.28; 411,18.36; 412,9.12.14.18.20; 413,27.28.29.32.33; 414,2.3.5.9.12.14; 415,17.21 etc.; travel, 486,20
kinêsis, change, 394,4.5.6.7.9.11.12.14.15.19.20.21.23.24.25; 395,1.4.12.15.19.23.27.28.30.31.33; 396,2.3.4.5.6; 397,1.3.7.8.9.11.13.14.15.16.19.20.21.22.27.29.34; 398,13; 401,3.8.12.13.16.23.34.35 etc.
kinêtikos, initiator of change, 436,14.17; 449,34.36; what changes, 425,19; source of change, 439,24.27; 440,2.14
koinônein, agree, 452,30; share, 461,17
koinônia, community, 499,8
koinos, common, 395,5.7.12; 396,19; 402,20.23.25.26.27; 403,18.25.29.36; 404,3.4; 412,4; 415,19; 416,1.2; 417,9; 422,31; 462,12; 467,6; 475,23; 476,20.26.30.32.33; 477,20.23; 494,16; 498,1.5; 507,13; general, 395,26; 440,25.30.33; 476,26.28; 480,23; 482,4; 490,4.10.32; universal, 417,5
koinotês, what is common, 452,27; 477,25; generality, 491,1.2
kôluein, prevent, 403,13.31.32; 443,6; 446,21.26; 455,24.29; 462,15; 464,1; 467,21; 468,3; 473,33; 478,1.34; 489,30; 494,29.35; 500,1; 508,6.8.11; 516,34; block, 470,28
kosmikos, cosmic, 467,36
kosmopoiia, making of the cosmos, 453,13

kosmos, cosmos, 453,16.20;
507,28.30.34; 508,2; 511,16.34.35;
universe, 464,3; 467,16; 487,20;
498,9; 507,34; 512,7
kouphotês, lightness, 488,8.10.12.16;
489,6
kratein, be victorious, 423,9;
overcome, 478,31; contest, 503,30
kreittôn, superior, 403,21; 407,9.19;
429,5; 461,1
krikôtos, circular, 500,17
krisis, recognition, 401,32
kubernan, govern, 464,13;
465,9.19.21;
kuklophorêtikos, with circular orbit,
482,12
kuklos, circle, 466,14; 470,20.34;
500,17.18.21.29.30.32; 511,32; 512,1
kulisis, rolling, 417,24
kuriôs, strictly, 395,31; 396,9; 404,31;
407,4; 412,18; 414,36; 415,1.2;
416,17.18; 419,26; 425,2; 427,4;
430,24; 439,30; 458,5; 471,10;
472,9; 474,7; 500,23.32;
501,19.20.21.22.23.26.32

laburinthos, labyrinth, 470,24.29.33
lambanein, consider, 396,5; assume,
493,28; 497,22; 510,20; gain,
498,22; attain, 497,32; 498,7.25;
500,16; 508,10.12; 513,5; derive,
477,9; 502,13; take, 399,23; 401,22;
407,8; 410,24; 415,29; 430,14;
433,8.17; 435,21; 437,26; 441,5;
442,12 etc.; (Pass.), be given,
491,30; 494,28; 495,4.23.24; 497,24;
498,7.15; 499,22; 500,33.35; 504,26;
506,19.22.36; 508,7.15; 509,9;
518,15
lamprôs, evidently, 514,8
lêmma, assumption, 401,1.7; 402,9;
413,15; presupposition, 427,26;
premiss, 416,11; lemma, 402,19;
404,13
leptomerês, of rare parts, 432,30
leukos, white, 398,21.22.23; 400,18;
406,24.28.32; 410,34; 411,2.11;
413,24.28.29; 415,33.34; 425,30;
444,23; 452,7; 473,30
leukotês, being white, 412,17;
whiteness, 413,29.30; 430,9; 451,33
lexis, passage, phrase, 395,21;
407,27.36; 409,26; 422,19; 430,2;
440,13; text, 429,28; 436,13;
443,13; 450,33; 481,29; 449,6;
reading, 427,35; language, 439,22;
words, 416,31; 437,6; 454,17; 495,3;
ordinary use, 401,30
logikos, rational, 430,19; verbal,
440,18.22.24.26.29
logos, account, 394,13.21; 395,12.18;
397,22.35; 403,22; 406,26; 409,8;
410,27; 413,8; 417,32; 425,15.17;
426,10; 427,30; 431,2.18;
433,3.9.11; 437,21; 443,7; 445,6.11
etc.; reason, 446,2; word, 440,22;
definition, 394,15; 397,8; 418,35;
423,30.32.35; 439,34; 449,33;
502,30; argument, 420,19; 427,6;
436,7; 440,23.26; 442,14; 445,4;
449,20; 463,3; 467,1.4.35; 469,2;
477,8.11.17.20; 484,2; 487,5.15;
488,25; 491,7; 495,12; 507,1;
510,23.30; 512,11; 514,32;
515,7.10.21.22.24.30; 516,3.7;
517,3.6; 518,2; discussion, 397,4.6;
404,2; 417,2; 421,3; 422,34; 424,7;
432,5; 465,17; 469,25; 475,32;
476,23; 482,10.14.21; 487,14;
498,20; 503,5.12; 513,7; 515,10;
(***hoi logoi***) lecture, 443,28; 454,20;
503,12
loidoria, abuse, 438,10.11
luein, remove, 410,17; 445,9; 492,19;
refute, 517,6; 518,2; rebut,
445,16.18; 447,20; solve, 418,11;
421,9; 422,12; 439,15; 480,19;
resolve, 478,35
lusis, rebuttal, 445,20; solution,
408,31; 506,19

manthanein, learn, 413,25; 421,2;
444,30.31; 445,25.34; 446,18.29;
447,8.13.15.16.19.20;
448,2.3.4.5.6.7.10.13.15.20.24;
449,9.13; 467,30; 469,8; 470,1
marturein, bear witness, 411,15;
420,24.32; evidence, 518,14
matên, pointless, 426,28.30.31.34;
464,20.23.24; 492,16
(***ta***) ***mathêmata***, mathematics,
492,5; 511,4; 512,36
mathêmatikos, learning process,
446,15; mathematical, 467,10;

468,10; 469,24; 475,24.27.30; 476,10; 477,20.23; 512,19; 517,5; (*ta mathêmatika*), mathematical objects, 453,7; 476,20; 512,23.33; mathematics, 477,10; (*ho mathêmatikos*), mathematician, 466,10; 510,31.32; 512,4
mathêsis, learning, 417,23; 442,26; 444,18.28; 445,31; 446,26; 447,12; 448,10.12.14.19.22.23.29.31; 449,7.8.9.13; study, 510,28
megethos, magnitude, 451,13.15.23; 452,1.4.37; 454,31; 455,26.29; 458,24.27.29.31.32; 459,20.23.24.26.33.34.35; 460,2.3.4.8; 461,9.30; 462,2.5.6.7.9.13.19; 463,4.20.22.23 etc.; size, 470,21.23
meiôsis, diminution, 397,33; 403,21; 416,4.5; 422,28; 431,32; decrease, 407,15; 468,16; 497,9
mêkhanêma, system, 515,29
melania, blackness, 413,31; 451,33
melansis, becoming black, 397,31; 407,16
melas, black, 406,24.27.28.33; 410,34; 411,2.11; 413,30.32.33; 414,1; 415,33.34; 444,23
menein, remain, 399,18; 400,36; 406,10; 408,20.22.24; 414,4.6; 416,36; 418,27; 427,18.24.30; 433,20.34; 434,30; 435,11; 438,1; 442,19; 446,19; 450,21.24; 478,10; 482,26.27 etc.
merikos, particular, 489,28; 490,21; 501,20
merizomenon, portion, 496,19
meros, part, 410,36; 417,11.12; 420,32; 453,22; 454,2.4; 474,12.14.21.22.23.24; 475,19; 483,16.20.34; 484,22.24 etc.
mesos, intermediate, 425,1; 475,23; mean, 434,17.21; 446,7; middle, 469,4; 487,21.34.35; 501,30.33; 502,2
merizein, separate, 449,1.3; divide, 423,12; (Pass.) have parts, 493,14
metabainein, move over, 409,4; turn to, 428,17; 439,4; 444,14; 465,27
metaballein, change, 398,12; 423,6; move, 409,18; suffer transformation, 443,1; (Pass.) be transformed, 398,22; 399,13; 402,2.32.33; 403,1; 404,29; 405,1; 408,23; 409,3.5.17; 410,6.9.21.22.26.34 etc.
metabasis, moving about, 422,29
metabatikos, spatial, 420,32; transitional, 421,17.31; 422,4; 423,4; transition, 422,7
metablêtikos, being transformed, 481,27
metabolê, transformation, 395,23.25.26.27.28.29.31; 402,7.32.33; 404,30; 406,9.18; 407,13.16.22.32; 408,2.4.5.6.11.22.24.25.27.34; 409,7.8.11 etc.; change, 403,22; 431,24
metalambanein, share, 405,22; substitute, 416,27; convert, 437,28
metalêpsis, exchange, 437,23; version, 450,30; participating, 502,3
metapherein, cause change of place, 423,7; transfer, 502,16.17
metastasis, transition, 412,13
(*to*) *metaxu*, intermediate, 408,35; 443,36; 444,1; 452,32; 458,25.35; 459,3; 471,4; 484,12; 514,34
metekhein, partake, 398,13; 406,21.22.23; 432,37; 433,17; 434,9; 453,34; 454,34.36; 455,7; 465,17; 477,26.34; 497,26; 502,1; share, 405,11; participate, 430,29; 433,5.11; 461,17; 497,21; 501,5.6; possess, 412,18; have a share, 421,11
meterkhesthai, move on from, 472,10
methexis, relationship, 398,14; participation, 430,28; 433,3; partaking, 477,32
methistanai, transform, 411,31; turn from, 412,24; shift position, 412,27
methodos, method, 476,27
mekhê, participation, 454,14
migma, mixture, 460,8.24.27.35; 485,22.30; 486,15.23.24.33.36
mimnêskein, call mind, 437,7; refer, 467,4; recall, 475,22; 515,12
mixis, mixture, 460,28.29.31; 488,15
mnêmoneuein, recall, 401,10; 443,32; 498,18; make reference, 432,34
monadikos, unitary, 505,33; 506,31; monadic, 506,5

monas, unit, 421,10; 452,4;
  454,9.24.33.34; 455,28;
  456,22.23.25.26.31; 457,9.17.20;
  467,10; 471,4; 474,15; 477,36;
  499,13.35; 504,29; 505,19.21.23.35;
  506,12; 516,34.36; monad, 458,8.10;
  505,11
monê, rest, 414,7; 486,17.21; 487,16;
  488,9
monimos, stable, 432,8
morion, part, 395,11; 396,19; 419,11;
  454,33; 459,29; 466,5; 482,23.27;
  483,10.11.30.31.35; 485,19;
  487,5.7.28; 491,17; 498,23; 503,8;
  505,7; 516,15
morphê, form, 398,10; shape, 407,10
morphousthai, have a shape, 497,26

neos, modern, 420,11
neikos, strife, 459,12; 464,12; 465,7.12
neuron, vein, 460,16; sinew, 484,25
noein, intellect, 420,20.21;
  understand, 517,16; think, 453,7;
  461,4; 517,8.11.14.18.28.29.30.33;
  conceive, 430,25; ***to noein***,
  intellection, 420,20.23
noeros, intellective, 404,23; 405,29;
  419,6; 435,2
noêsis, intellect, 461,25.26; thought,
  467,6; 517,1.3.4.7.31; 518,14.15; ***en
  noêsei***, the conceived, 515,16
noêtos, intelligible, 400,15; 453,18;
  461,12.19; 469,23; 475,23;
  476,10.21; 487,20.22;
  503,16.18.19.25.32; object of
  intellect, 475,30; conceived, 476,31;
  to noêtos, intelligible world, 476,6
nomos, convention, 512,29
nous, intelligence, 405,33.34.37;
  406,3; intellect, 420,25.32;
  421,1.24.31; 422,8; 451,7;
  459,13.15; 460,1.4.5.14; 464,12;
  465,6.12; reason, 420,30
(***to***) ***nun***, instant, 468,31

oikeios, special, 396,13; proper,
  483,14.28; 485,23; 487,14;
  appropriate, 413,1; 423,18; 462,20;
  469,16.22.23.25; 476,29; 477,2
oikeiotês, affinity, 485,26
(***to***) ***on***, being, 397,26; 398,4; 402,26;
  404,19.22; 405,1.7.9.13.25.27;
  406,18; 407,23.30.32;
  408,6.12.14.16.17; 412,29;
  430,14.16.23.24.27; 432,17;
  501,20.22; 502,4; 503,22; has being,
  420,26; thing, 400,25; 401,2;
  405,4.9; 432,7; 454,22; 456,6.10;
  471,2; 480,26.34; 481,20; 487,12;
  499,30; 501,23; 503,17; what exists,
  398,2; 409,2; 416,18; 423,34;
  431,20.34; 468,2; 480,32; 499,11;
  502,5; 517,13; reality, 399,9.10.12;
  entity, 398,17; 451,27; existence,
  416,6; 430,20; 433,16; 434,11.17;
  452,24; actual, 511,29
oneiropolein, dream, 517,9
onkos, mass, 467,18
onoma, word, 400,36; 401,24; 406,15;
  414,37; 415,12; 441,6.19; 474,5;
  500,15; 505,17; name, 403,25.26;
  406,14; 415,19.31.33.35; 416,1.2;
  421,20.33; 439,9; 447,23; 488,35;
  (***kat' onoma***), verbal, 428,10
onomazein, call, 401,26; 416,4; 422,3;
  442,6; 482,34; 500,18; name,
  407,17; 505,19
opsis, sight, 425,21.24; iris, 444,24;
  (***pros opsin***), before our eyes,
  513,12
organikos, organic, 414,26; 482,36
organon, organ, 426,7; tool, 494,2
ouranios, heavenly, 419,7; 505,37;
  (***ta ourania***) heavenly bodies,
  419,24.26.28; 421,6; 435,3.17
ouranos, heaven, 394,24; 421,13.19;
  453,14.18.19.20.23; 455,18;
  467,14.27.36.37; 468,1;
  507,30.32.35; 517,6.12; universe,
  507,4
ousia, essence, 397,21; 398,5.12.13;
  419,7.8; 420,25; 425,25; 440,27;
  444,34.35; 445,2.3; 478,8.11;
  substance, 402,15.18;
  403,14.19.27.34; 404,29; 405,2;
  406,20.21.29.31; 407,5.33;
  408,4.8.17; 409,1.2.10 etc.
ousiôdês, substantial, 398,5
ousiôsis, coming in existence, 433,17
ousiousthai, come in being, 433,37
oxunein, give an oxytone accent,
  443,10; 490,17

Indexes 189

pakhumerês, of dense parts, 432,30; dense, 436,10.11
palaios, ancient, 420,10; 424,35; 499,1; old, 420,12; earlier philosopher, 397,14
pan, universe, 432,30; 460,33; 464,5; 467,2; 469,1; 480,31; 483,32; 489,24; 441,19; 507,19.22; 509,12; 511,31.35; 512,17.20.25; 515,8; 516,15; totality, 517,38; universal, 460,34
panspermia, universal seed-bed, 459,26
parabolê, comparison, 482,36
paradeigma, example, 415,14.22; 416,11; 417,23; 423,22.28; 425,16; 440,34; 447,1.23; 448,11; 453,13; 476,33; 481,28.33; 486,6; illustration, 497,20
paradeigmatikos, paradigmatic, 406,13
paradeiknunai, show, 432,5; 492,20
paradeixis, argument, 438,3
paradekhesthai, accept, 462,20
paradiazeuktikos, alternative, 478,1
paradidonai, give account of, 395,7; set out, 504,27; transmit, 406,9; hand down, 395,14
paragôgê, mistake, 453,6
paragraphê, marginal note, 450,32
paragraphein, set out, 461,16
paragumnein, lay bare, 465,36
paraitein, repudiate, 427,35; hesitate, 451,6; excuse, 478,25; beg to differ, 483,1
parakolouthein, accompany, 454,12
paralambanein, include, 394,7.13.15.17; 396,11; 397,8; 400,2; 449,34; take as equivalent, 510,7; add, 400,23; 407,29; accept, 428,30
paraleipein, omit, 427,28; 441,22.38
parallagê, lack of fit, 499,8; (*ekhein parallagên*), illustrate, 517,19
parallattein, alter, 517,25
parallêlogrammon, parallelogram, 457,4.5
paralogismos, fallacy, 463,3; 502,18
paramutheisthai, assuage, 510,30; explain, 444,16
paraplêrôma, complement, 457,3
paraskeuê, preparedness, 400,28

paratasis, extension, 461,25; 494,15.30; is extended, 493,8
paratithenai, add, 400,11; set out, 432,20; 476,33; insert, 415,23; quote, 433,14; 437,1; include, 477,22
parauxanein, make greater, 456,22; increase, 456,36
parauxêsis, increase, 517,23
pareinai, be present, 399,5.6; 431,10; 433,14; 434,8.24; 442,22; 445,27
parêkein, set aside, 395,17; omit, 409,9; 441,29; 444,7; 468,8; 477,29; 478,17; 484,21; 488,13; 490,1
parekhein, permit, 396,24; make, 496,12; 503,24; confer, 396,32; serve, 465,10; supply, 456,6.10; provide, 398,34; 432,34; 487,25; 503,27; 506,35
parelêluthos, past, 494,21
paremphasis, exhibition, 445,2
parergôs, as a side issue, 452,24; incidentally, 469,19
paroimia, proverb, 502,12
parônumos, paronymous, 505,17
parousia, presence, 442,22
parüpostanai, be a by-product, 429,21
paskhein, be passive, 396,16; 439,12.14.18; be acted on, 436,6.17; 440,7.12; 444,10.21; be affected, 419,13.24.25.33.34; 426,3.29; 434,32; 439,23; 440,11; 441,15.16.31.33.37 etc.
pathêsis, suffering, 401,18; passion, 439,13.14; affection, 441,7.16.33.37; 442,3.7.12.26; 444,29; 446,26; 448,31; 449,3.11
pathêtikos, passive, 401,11.12.17.24.29; 449,31; affection, 449,30; affective, 451,32
pathos, passivity, 404,36; 405,2; being acted on, 422,32; experience, 411,12.13; affection, 411,5; 412,26; 415,18; 419,31.34.35; 426,4.28; 439,30; 440,6; 441,9; 451,23.32; 452,7.11.13.14.15; 461,22; 473,2; being affected, 410,26
pattalos, peg, 485,31
peithein, persuade, 405,31; 466,30
pêlos, mortar, 427,10.14
peperasmenos, limited, 451,15.26.34; 452,1.2.3.8.9.14; 453,37;

458,29.32.33; 459,9.32.34; 463,5.25.26.29; 466,32.34.35; 467,2; 470,31; 474,27.30.31 etc.
perainein, traverse, 425,21; limit, 395,10.19; 453,34; 454,16; 455,23; 456,6.12; 466,32.37; 467,3; 484,20; 488,20.34.37; 492,34; 495,4; 497,27.29; 498,9; 499,14; 502,25; 515,16; 516,1
peras, end, 409,22.24; 420,8; extreme, 512,1; limit, 425,20; 429,10; 451,29; 452,4.6; 454,24.35; 462,24 etc.
peratoun, limit, 503,1
peratôtikos, providing limit, 456,16
periekhein, encompass, 404,4; surround, 411,32; contain, 432,7; 453,20; 462,9; 464,13; 465,9.19.20; 485,24.25; 502,14.16.28.35; 503,1.10.13.19.22.25.27; 504,2.3.6.7.18.19; 505,27; 507,7.8; 513,2.3.4
periekhon, container, 482,18.19; 504,14
periektikos, containing, 515,2
perigraphê, circumscription, 464,8
perilambanein, include, 496,12.17.21; 497,5.7; 500,28; embrace, 518,17
periodos, cycle, 499,36
periokhê, content, 430,26
periorismos, boundary, 504,11
(*ek*) *periousia*, superfluously, 448,23; 480,25
periphereia, circumference, 466,14.15.16
(*kata*) *peristrophê*, as we go round, 500,19
perithesis, placing around, 457,15; surrounding, 457,17; putting around, 457,22.24; 458,10; enclosure, 458,3.9
peritithenai, place round, 457,2.8; add (around), 457,8.17; put round, 457,20.23; 458,1.7.8; set about, 518,17
peritrepein, turn in the opposite direction, 512,10
perittos, odd, 429,11; 456,6.12.15.16.24.25.28; 457,1.7.10.14.23; 458,2.4; 492,10.11; superfluous, 441,33
perix, extreme, 469,5

phainesthai, be seen, 427,7; 443,1; 468,33; appear, 460,33; 465,30; 466,32; 468,12; 494,10; be visible, 443,27
phainomenon, observable, 396,7; apparent, 499,7; superficial, 487,18; (***kata phainomenon***), from observation, 486,30; 487,12; at a superficial level, 498,34; 499,28; superficially, 503,21
phantasia, imagination, 465,32; 467,6; 477,9.28; 506,10; 517,3.12
phantazesthai, imagine, 467,13; 517,8.14.17.25
philia, love, 459,13; 464,12; 465,6.12
philologein, (do) philology, 407,4
philos, friend, 405,8; 412,23
philosophia, philosophy, 451,7; 458,20
philosophos, philosopher, 404,16
phleps, sinew, 460,16
phobos, fear, 510,29
phônê, word, 400,26; voice, 425,34.36; 426,1.2.3.5; 470,7; 471,30.34; 472,5.17.18.19.20.21.22.23; expression, 404,15; sound, 516,24.28
phônêtikos, vocal, 426,7
phora, place, 407,8; travel, 411,3; 413,4; 415,26; 416,1; 417,25; 422,29; 432,29; spatial motion, 420,31
phôs, light, 429,14
phronêsis, thought, 405,31; intelligence, 406,5
phronein, think, 405,32
phthartikos, destructive, 484,13.17
phthartos, perishing, 412,11; destructible 467,21.24; 482,14
phtheirein, destroy, 399,13; 465,1.2.3; 479,5; 480,4; (Pass.), perish, 398,35; 427,1; 467,22; 481,7.8; cease to be, 410,12.15.36.37; 411,2.26.28; 412,17.19; 434,29; 480,14.15; 494,20.21.30.35; 506,7.8.32
phthisis, decay, 416,4
phthora, ceasing to be, 397,30; 403,20; 407,15.35; 408,4; 409,12; 410,14; 416,4.7; 417,6.8; 419,9; 435,18; 436,18; 480,13; 482,11; 515,14

phusikos, natural, 394,21; 395,6.9.18.19; 396,16; 404,27; 414,26; 419,28; 420,23; 421,2.14.16.18; 422,3.28.33.34; 423,23; 426,27; 435,12.30; 450,4; 451,4.12.13.18.20.27; 452,25; 458,20; 465,17; 469,5; 472,14; 475,30.34; 476,24.32 etc.; (*ho phusikos*), natural scientist, 394,23; 394,5.18; 452,10.14; 465,25; 469,14.16.22.23.29; 471,3; 485,3; (*ta phusika*), natural things, 457,32; 418,2.6; 419,4; 422,7.13; 475,28; nature, 477,19
phusikôs, from a point of view of natural science, 478,5; naturally, 423,18; 434,33; 435,1.3.10.11; 436,8; 437,24; 438,6; 450,8; 478,10
phusiologia, study of nature, 452,23
phusiologos, natural scientist, 394,10; 397,2; naturalist, 420,10; 452,22; 461,24; 462,25; 464,18; 465,26; 469,18; 484,8; 507,22
phusis, nature, 394,4.5.6.7.9.11.12.13.20; 395,23; 397,7.17; 407,1; 415,6; 417,11; 422,17.34; 423,7.8.16; 425,3; 426,29; 432,27; 439,32 etc.
pisteuein, confirm, 428,21; 451,31; 477,8; 514,32; believe, trust, 465,34; 491,5; 517,1
pistis, evidence, 452,22; 461,15; ground, 465,23; confirmation, 466,7; 468,4
pithanos, persuasive, 441,38; (*to pithanon*) plausibility, 440,22
plastos, moulder, 445,28
plattein, mould, 445,29.34.35; contrive, 517,13
plêgê, beating, 426,4.5.8
plêmmelês, unruly, 422,6
pleonektêma, pretension, 502,18
plêrês, full, 501,31
plêroun, use up, 495,35
plêthos, many, 429,12; number, 460,1; 461,8; 468,20.26; 471,12.26; 478,24; 479,18.21.26; 484,26; 492,17; 493,18; multiplicity, 454,29.30.31; 455,7; 474,16; quantity, 458,26.27.28.29.32; 459,16.17.18.23.25.34; 460,3.5; 461,31.34; 462,1.3.10

plêttein, reverberate, 426,1.4.6
pleura, side, 456,21.27.34; 457,21; 466,17.18.19.23
pneuma, breath, 424,34
poiein, be active, 396,16; 439,12; posit, 464,11; bring about, 437,6; 494,29; 497,23; 507,17; conduct, 476,34; 482,20; 487,4; make, 403,23; 430,3; 437,28; 438,14.22; 439,4; 443,13; 455,9; 456,21; 457,6.21; 458,5.29; 459,9.16.20.24; 460,1.3.33; 462,5; 466,20.22; 469,25; 475,2; 477,33.34; 483,4; 494,4.6; 495,15; 496,10; 499,4; 502,30; 504,8; 512,2; 513,14; treat, 452,30; produce, 409,32; 438,19; 441,15.32; 456,2.35.36; 457,22; 481,26; 482,6; 511,10
poiêma, product, 441,8
poiêsis, making, 400,35; action, 439,13; 444,12; creation, 464,3; agency, 441,7.15.16.18.19.33.37; 442,1.3.4.5.11.25; 444,29; 446,26; 448,31; 447,3.11
poiêtikos, efficient, 394,3; 438,23; 459,12; 465,11; 513,11; 514,35; creative, 463,32; 464,11.14; agent, 441,4
poion, quality, 398,26; 402,33; 403,32.35; 404,3.30; 406,20.22.30.32; 407,8.16; 408,17; 413,3.24; 415,32; 438,21.26; 474,6; 501,4.5.6; (*kata poion*), qualitatively, 501,4; what, 468,13.14
poiotês, quality, 397,31.32; 400,17; 402,23; 403,21.27; 406,23.25; 407,16.17; 408,8; 411,23; 412,11.25; 415,35; 420,27; 424,36; 425,3; 434,3; 435,32; 438,21; 451,33; 481,21.22; 488,13; 512,24; (*kata poiotês*), qualitative, 395,30
poioun, agent, 415,13.21.22.23; 440,8.9.10; 441,16.19.32.33.37; 442,3.7.12.28; 443,28; 444,6.9.20.25.32; 445,23.24; 446,4.6.24; 448,32.33; 449,4.16; 463,9; efficient, 464,2; 513,14; creator, 464,4
polueidês, multiform, 438,3
poluthrulêtos, much discussed, 458,35

pompholugizein, bubble, 460,22
poson, quantity, 395,33.34;
    396,1.4.5.20; 402,33; 403,34;
    404,2.29; 405,2; 406,20.30.34;
    407,7.15; 408,18; 413,3.23; 416,1.3;
    417,28; 438,21; 452,1.2.8.10 etc.
posotês, quantity, 377,32; 402,23;
    403,20; 408,8; 412,12; 417,25;
    420,27; 438,21; (***kata posotês***),
    quantitative, 395,30
pragma, thing,
    402,6.10.12.17.22.30.31; 404,18.37;
    407,24.28; 451,12; 463,8.11.14.33;
    467,13; 468,13; 497,21.26.27.31;
    fact, 440,23; 442,19; 452,21;
    matter, 465,30; 476,29; 477,2;
    object, 517,7.8
pragmateia, investigation, 395,7.8;
    451,13.18; work, 408,19.28; 420,8;
    465,18; 492,4
pragmateuein, investigate, 451,17;
    452,16; study, 452,25
proairetikos, deliberate, 426,7
proballein, project, 400,31.34;
    406,15; propound, 513,8; introduce,
    487,14; (Pass.), profess, 492,9
problêma, problem, 439,5; 465,28;
    471,16; 515,10; questions, 452,18
proêgeisthai, come first, 395,6
proêgoumenôs, particularly, 507,35
proektithenai, put forward, 515,11
proemphasis, first manifestation,
    422,6
proüparkhein, precede, 405,22
proienai, go on, 401,3.13; 441,26;
    472,26; 510,10; arise, 440,22;
    proceed, 440,28.30; 450,26; 453,34;
    454,7; 455,1.11; 458,13; 504,33;
    507,27; 508,35.36
prokeimenon, subject-matter, 395,8;
    matter at hand, 404,34; 405,6;
    418,34; 469,25; 478,7; thesis, 467,32
prokopê, progress, 450,15; advance,
    505,25
prolambanein, assume first, 397,36;
    419,3; 424,18; 491,28; presuppose,
    483,27.34
prolêpsis, presupposition, 484,30
proodos, progress, 469,1; process,
    494,35; 506,33; 507,8.21; 508,24
prophasis, excuse, 502,18

pros hêmas, conventional, 489,14;
    subjective, 489,23
prosapodeiknunai, prove, 423,28
prosaptein, apply, 456,7
prosdiatithesthai, be predisposed,
    450,16
proseinai, be present, 450,13; 464,21;
    exist in, 476,7; reach, 500,24
prosekballein, extend, 511,10.12
prosêkein, be suitable, 407,11; 476,2;
    484,16; 501,8; be characteristic of,
    468,27
prosekhein, notice, 415,11
prosekhês, immediate, 421,26; 425,1;
    proximate, 435,35; 440,30;
    immediately, 495,11; relevant,
    440,25
prosekhôs, immediately, 421,5.7;
    438,18; next, 429,28; just, 442,11;
    485,16; 491,34
prosgignomenê, concomitant, 506,1
proskeisthai, be added, 441,31;
    495,12; be joined, 514,9
proslêpsis, additional assumption,
    493,1; 515,33
proslogizesthai, argue for, 465,5
prospoiein, suppose, 505,34
prosthesis, addition, 396,27.28;
    455,37; 456,18; 468,7.12; 469,28;
    470,36; 471,3.6.8; 477,13; 478,2.3;
    493,1.2;
    495,18.20.22.27.29.30.32.33; 496,5;
    497,10.29.31.34.35 etc.
prostithenai, add, 396,17; 400,3;
    403,26; 414,5; 417,32; 418,35;
    419,1; 422,15.25; 426,15; 449,29;
    450,33; 452,5; 454,2 etc.; propose,
    471,1
protasis, proposition, 511,22;
    premiss, 440,32
proteron, prior, 502,1
proümnein, remind, 422,13
proüparkhein, pre-exist, 441,1;
    461,12; exist before, 461,19
proüpokeimenon, pre-exists, 424,24
pseudos, false, 451,28.30
psileisthai, become weaponless,
    412,6.28
psukhê, soul, 397,16; 402,16; 403,2.6;
    405,31.35.37; 414,25.27; 415,3;
    417,24;
    421,3.5.6.14.17.18.19.21.23.29.33;

422,1.4; 423,4.16; 438,27.33; 440,27.28.30.31; 441,1; 445,32; 474,18; 476,6; 499,36
psukhein, cool, 410,5.8.9.12; 412,24.25; 419,23; (Pass.) become cool, 418,23
psukhikos, psychic, 422,8
ptaisma, error, 481,34
pur, fire, 396,15; 399,1; 423,19; 450,18; 453,20; 458,34; 460,13.14.21; 465,14; 471,3; 478,31.32.33.34.35; 479,3.4.5; 480,2.27.33; 482,24; 484,9; 486,28
pureisthai, be heated, 423,19

rhabdos, rod, 409,22; staff, 467,28.32
rhein, flow, 406,10
rhêma, statement, 461,16
rhêsis, word, 400,11; 431,6; statement, 432,20
rhêton, sentence, 399,31; statement, 429,32
rheustos, fluid, 503,15
rhopê, inclination, 469,5; 488,8; tendency, 489,6

saleuein, infringe, 444,3
saphês, clear, 443,13; 478,25; 484,21; 495,12.14; 496,12; clearly, 495,31
saphôs, plainly, 430,12.21.28; clearly, 464,26; 476,3; 517,7
sarx, flesh, 460,16.23.27; 482,31; 484,25; 514,16
selênê, moon, 419,33; (***ta hupo selênên***), the sublunary (affairs), 419,18.19; (***hupo selênê***), sublunary, 482,5.15.20
sêmainesthai, mean, 403,7; 416,4; 446,20; 463,7; 496,18; refer, 475,30; signify, 403,25; 405,29; 463,8; 469,31
sêmainomenon, signification, 405,10.13.17.25; 469,32; sense, 451,3; 470,1.14.16.22.23.26.36
sêmeion, sign, 396,7; 411,19.23; 452,19; 456,16; point, 454,24; 466,19.24; 468,14.31; 470,5; 516,27.36
sêmeioun, note, 395,32
semnotês, dignity, 502,13
skepsis, inquiry, 487,13
skhêma, form, 399,15; 401,25; shape, 411,33; 445,32; 457,2; 459,27.28; 462,6.8.13; 464,9; 470,19; 470,35; figure, 418,8; 424,13; 425,5; 457,22; 511,28
skhêmatikos, diagrammatic, 457,16
skhêmatismos, configuration, 398,16
skhêmatizein, draw, 457,19
skhêmatographein, use diagrams, 457,17
skhesis, relationship, 396,6; 409,32; 419,12; 437,6.11; 439,7.8; relation, 448,33; 449,4; 488,5; relative position, 409,19.20.21; (***kata skhesis***), relational, 489,23
skholê, comment, 461,15
sklêros, harsh, 416,23
skopein, look at, 429,30.35; 476,12; examine, 469,29; observe, 482,1; (***to***) ***skopeisthai***, consideration, 421,20
skotos, darkness, 429,14
smikrotês, smallness, 494,4
sôma, body, 394,21; 395,9.13; 396,16; 397,4.17; 398,11.21; 411,6.14; 414,27; 417,24; 419,8; 421,5.9.16; 422,2; 423,6; 424,9; 426,2.29; 434,34; 435,31.32; 438,10.27; 451,13.14.18.19.20.22 etc.
sômatikos, bodily, 394,21; 420,10; 435,31; 461,31; 465,36; embodied, 506,24; of body, 467,18
sômatoeidês, bodily, 456,13
sophos, wise, 461,12; 510,18.26
sôphrosunê, prudence, 476,4.6
sôzein, preserve, 478,30; 493,23
sperma, seed, 413,22
speudein, seek, 433,22.37; speed, 434,12
sphaira, sphere, 420,31
sphendônê, bezel, 470,19; 500,16
spinthêr, spark, 482,24; 486,28
stadion, mile, 517,14
stasis, rest, 404,24; constancy, 430,15.16.20; stability, 432,25; 493,1; (***en stasei***), static, 414,7
stereisthai, be deprived of, 395,13; 397,4; 406,22; 434,12; 484,38
stereos, solid, 424,35; 454,25
sterêtikos, privative, 407,20; 428,28.31; 429,3.7.19; 434,9.11; 513,33; 514,23
sterêsis, privation,

406,29.31.32.33.35; 407,5.10.20.33; 412,1.5; 428,20; 429,5.6.20; 432,7.10.11.12; 434,14.22; 451,29; 513,17.18.19.20.22.23.24.26.28.29. 30.34; 514,1.2.5.6.11.18.19.20.21.32
stêrigmos, support, 487,24
stenokhôria, lack of space, 511,13
stigmê, point, 451,32; 452,2
stizein, punctuate, 399,19; 400,7
stoikheiôdês, like an element, 463,24; as an element, 464,10
stoikheion, element, 405,21; 419,20; 425,3; 454,15; 455,4; 458,24.30.33.35; 459,2.4.6.16.20.22.31; 460,1; 464,3; 466,31 etc.; numeral, 457,18
skhazein, aim at, 432,4
strophê, rotation, 500,22
sullambanein, combine, 434,7; contain, 501,23
sullêpsis, combination, 434,7
sullogismos, syllogism, 416,10.11; 425,4; argument, 430,13; 477,22
sullogizesthai, argue, 424,20; 425,6; 476,27; 501,10
sumbainein, result (in), 441,22; 474,32; 486,8; 491,8; 504,1; come about, 441,24; 500,3; 504,12; 505,31; 506,11
sumbebêkenai, happen accidentally, 517,28.29.30; be a coincidence, 517,2
sumbebêkos, attribute, 402,17; 425,23.24.25.35; 431,3; 439,8; 441,13; 452,29.30; 458,23; 459,33; 468,25; 469,7.9.11.13; 471,17.18.19.21.24.25.32.35; 472,1.9.10.26.28.31; 473,7.8.13.28; 474,6; 475,4.6.9.13; 476,2.4.9.15; 477,4; 488,3; 498,17; 513,16; (*kata sumbebêkos*), incidental, 409,5.17.18.21.23; 462,28; 464,25; attributive, 471,33.35; 472,9
sumbolê, point of junction, 494,8
summetria, balance, 424,35
summignusthai, be mixed with, 497,17
sumparerkhesthai, pass by together, 411,4
sumparateinein, extend, 461,26
sumparousia, co-presence, 419,14
sumperainein, draw conclusion, 432,31; 485,20; infer, 436,25; sum up, 449,20; go round, 500,30
sumperasma, conclusion, 427,28
sumphônia, agreement, 404,21
sumplêrôtikos, constructive, 403,7; completion, 513,31.33
sumplêroun, complete, 396,12; 464,34; constitute, 449,2; (Pass.), become whole, 502,2
sumplokê, involvement, 446,10
sumpnoia, joint breath, 461,17
sunairein, involve, 437,17; unify, 487,27
sunêrêmenos, complex, 461,1.3
sunairesis, complexity, 461,2
sunaptein, join, 396,20; 495,10; 500,3
sunarithmein, count, 403,9
sunarmottein, be compatible, 400,16
sunêgoria, agreement, 412,34
suneinai, be together, 393,7; be present with, 493,30
sunekheia, continuity, 414,37; 514,24.25.26.27
sunekhês, continuous, 394,14.15.16.18; 395,4.17.20.33.34; 396,3.4.8.10.17.19.22.23.25.28.29.31. 34 etc.
sunêmmenon, conditional proposition, 484,34; 490,1; 515,33
sunemphasis, reference, 411,14
sunêtheia, common usage, 401,24; use, 500,15
sunêthês, characteristic, 461,24; customary, 491,27
sungramma, book, 460,26
sunistanai, confirm, 402,10.19; support, 463,15; constitute, 433,27
sunkatabainein, conform, 462,2
sunkeisthai, be composed of, 396,26; 419,20; 474,16; 484,24.26; 492,5.11; 504,34; 505,2.7.20.21; 507,11.13; 518,26; consist of, 454,9; 485,9; 488,30; 504,35; 505,15.21; 509,31; 510,11
sunkhôrein, accept, 405,34; 406,2; 443,8; 510,26; admit, 412,15.16
sunkhrasthai, use, 413,16
sunkhusis, muddle, 420,19
sunkineisthai, be changed together with, 409,21
sunklôthein, weave together, 502,10.11

sunkrasis, composition, 398,15; conjunction, 419,12
sunkrisis, association, 396,12; composition, 420,12
sunneusis, concentration, 432,30
sunnoein, notice, 411,8; understand, 461,7
sunodeuein, travel with, 419,33
sunokhê, fusion, 501,30
sunônumos, univocally (named), 403,12.27.28.31.32; 415,27; signified by common name, 403,24
sunousia, conference, 454,18
suntelein, contribute, 397,35; 398,1; 488,3
sunthesis, combination, 458,3; 499,35; putting together, 495,30
sunthetos, compound, 397,17; 398,9; 459,6; 463,28.30.31; 473,27; 474,18.19; 477,1.5; 478,8.12.13.14.15.21.29; 479,18.30; 480,9; 482,3.5.29.30.33.36; 483,2; 513,15.21; 514,14; composite, 481,25; 497,21; 502,23.25.27; 503,6.7; complex, 399,25; 488,4
suntithenai, put together, 494,5.7; 507,17
suntomia, short-cut, 471,18
sunüparkhein, co-exist, 481,11; be present in, 512,29
sustasis, constitution, 483,32
sustoikhia, column, 428,27.31; 429,3.5.8; 439,22; 458,15
sustoikhos, composed of the same elements, 419,30

ta hôsautôs ekhonta, constant, 451,5
tagathon, the good, 420,14
tattein, class, 403,30; 470,34
(to) tauton, identity, 404,20
tautotês, identity, 404,24; 424,19; 425,26.28; self-identity, 435,16
taxis, order, 398,1; 420,30; 452,18; 460,13; 461,2; 465,27; 477,6; 503,27; role, 406,27; status, 421,10
têde, HERE, 406,13; 441,1; 503,24.29
teinein, turn towards, 421,3
tekhnê, skill, 404,5; 450,25; 494,1; craft, 422,30.33
tekhnês, product, 463,22
tekhnikos, from a craft, 423,22; artificial, 450,4.36; manufactured, 501,17
tekhnikôs, by a craft, 423,17; artificially, 450,8
tekhnitês, craftsman, 494,2
tekmêrion, example, 434,34; 435,1.17
teleios, complete, 398,17; 400,24.28; 406,34; 407,7; 414,23.24.34; 415,1.3.6.7.9.23; 416,16.35; 417,17; 422,14; 426,20; 428,8; 432,7.13.15.17; 434,22; 438,17.23; 440,4; 499,15; 501,12.13.27.29.32; 502,1.4.24; 515,21; perfect, 450,9.23.24
teleiotês, completion, 414,22.24.27.31.32; 416,29; 417,14.20.22; 433,22; 438,26
teleutaios, last, 447,18; final, 466,37
teleutan, make complete, 414,35; end, 442,38; 443,5.7.9; terminate, 443,26; 464,34
teleutê, end, 433,33; 464,34; 491,29
telikos, final, 464,3.11.14; 465,11; 513,11
telos, end, 399,7; 446,11; 463,9.35; 464,7; 501,13.30.33; 502,2; 513,13; goal, 441,8.9.10
tetragônos, square, 429,8; 456,19.20.24.26.27.28.29.30.31.32; 457,2.7.8.9.21
theologos, theologian, 461,24
theôrein, see, 400,10.22; 423,12; 453,10; 461,27; 468,7; 481,20.24; 493,5; 499,13; 500,7; 515,5; observe, 400,17; 405,3.5; 471,1; study, 413,20; view, 400,37; 401,1; 402,18.32; 403,11.32.35; 404,28; 405,21.26; 406,5.8; 412,9; 413,1; 417,8; 429,19; 432,18; 434,5.12; 440,1; 501,30; 502,35; 503,6; 513,24; 514,22; *ho theôroun*, thinker, 512,21
theôrêma, theorem, 397,22; 510,19; 511,22.30; 512,18
theôria, study, 413,15; 468,19; 510,31; inquiry, 452,20.21
theos, god, 397,16; 465,16.17; 499,37
thermotês, heat, 430,9; 479,25; heating, 451,34
thesis, position, 411,31.33; 447,28; 454,24; 516,36; view, 441,34

(**to**) **ti**, essence, 397,34; something, 398,14
timan (v.), value, honour, 495,15; 499,37
tithenai, take, 479,15; 485,4; posit, 453,27; 459,34; 461,31.34; 464,19; 465,5.15.30; 491,7; 499,35; 512,13; 513,5; 515,18.19.20; 517,4; 518,2; treat, 405.25; 452,24; 462,23.27; 467,34; 468,25; 481,20; 483,2; place, 395,34; 396,1; 401,15; 428,22.26; 429,30.34.35; 431,26.29.34; 432,16.26; 444,5.6; 453,6.19.27; 455,13; 458,16; 499,10.31; class, 406,9
tmêma, section, 441,21.25.29.36; 442,2.11.24; 444,7; 484,22; 492,34; 498,12; 518,27; segment, 467,11; 471,7; 494,3; 498,27; 499,24; 514,26
to dioti, explanation, 486,4
to einai, essence, 394,5; existence, 397,13; 468,5; 484,20; 494,25; 508,13; 511,26; 512,29; being, 420,33; 421,28; 422,9; 448,1; 451,6; 464,9; 465,24; 471,13; 474,20; 492,24.27.28; 493,6.28; 494,17.33; 495,1.6.13.16; 497,15; 501,37; 506,6.8.30; 513,23; 515,18; 518,16
to hoti, fact, 484,4.5
to hou heneka, for the sake of which, 463,35; that which is aimed at, 513,13
to kath' hautên einai, self-subsistence, 490,33
to kath' hekaston, 402,21; 412,10
to pros ti, related term, 394,8; 409,32; 437,6; relative (term), 401,5; 409,2.12.14.15.27; 413,8; 437,32; relation, 396,2; 398,26; 401,14.30.31.33; 403,11.16.17.34; 425,32; 437,1.8.11.12.17.31; 516,9.12; thing related, 439,7; relational, 396,5; 401,9; relata, 437,10; 439,7
tode, particular thing, 398,3; 399,22.35; 400,4.7; 404,2; 473,26; definite particular, 456,15; substance, 438,30; 'this', 474,17.18.19; 490,12
toionde, quality, 398,3; 438,30
tosonde, quantity, 398,3; 399,22.36; 400,4.7; 438,30
tomê, division, 455,27.29; 466,10.12.16.26; 468,29; 519,27; cut, 396,24.28; 454,6; 455,34; 456,1.2.9; 507,14; 511,2.5; 514,15; section, 493,34; 494,5; cutting, 491,33; 492,33; 494,10; 502,31; 507,12; 511,17; 512,35; (**koinê**), intersection, 512,1; **topikos**, of place, 398,13; 419,9; 438,28.34; 484,31; 490,24; 510,7; in place, 490,23
topos, place, 394,22; 395,1.3.4.10.13.16; 397,1.4.9.10.11.18.33; 402,23; 403,1.21; 404,30; 406,34; 407,17; 408,30; 409,19; 410,34; 411,1.3.23.32; 412,1.3.8.23.32; 415,36; 420,28; 422,28; 423,6.8; 428,4; 435,21; 438,24.25.27.28.33; 448,13; 453,24 etc.
tososde, quantity, 514,27.29
trigônon, triangle, 466,17.18.20.22.25; 511,33.34; 512,2
tropos, way, 396,15; 429,32; 434,23; 449,5; 459,35; 467,30; 470,8; 495,21; 497,35; 499,23; 513,4; mode, 424,20

xurrhoia, conflux, 461,17

zêtein, inquire, 395,2.5; 408,5.15; 413,10; 417,1.11; 424,32.33; 439,5; 465,28; 470,15; 471,18; 472,10; 476,1; 511,30; (**to zêtein**), inquiry, 487,12
zêtêsis, inquiry, 420,9; 465,27; 469,17; 471,18; 475,20.30; 476,8.17.24.32.34
zôê, life, 405,31.34.37; 406,5; 414,26; 421,11.12
zôon, picture, 403,15; animal, 417,13; 422,29; 423,4; 430,19; 463,22; 470,10.11; living being, 421,13

Subject Index

Academy, 411,36.37
actualization of the complete, 415,8 ff.
Alexander (the name), 403,14
Alexander of Aphrodisias, 395,33;
  396,20; 399,19; 400,1; 403,13.23;
  407,21.36; 409,25; 414,16.29;
  416,27.31; 419,26; 420,13.18;
  422,22.25; 423,14.20; 426,6;
  427,34; 428,17; 429,23;
  431,7.27.33; 434,35; 436,19.27.36;
  437,6.9.23.31; 440,28.34;
  443,10.18.34; 446,13; 448,21;
  449,4; 450,30.36; 454,19; 457,13;
  459,5; 463,18; 467,1; 469,10;
  470,30; 472,36; 475,33; 478,18;
  483,1; 489,21; 490,18.20; 495,8;
  501,2; 511,30; 516,29; 517,17;
  518,8
analogy with building process, 426,20 ff.
analysis of Aristotle's claim on
  change, 397,28 ff.
  axiom arrived at by division,
  398,3 ff.
Anaxagoras, 452,2; 458,26; 459,13.17;
  460,4.6.7.11;
  461,1.6.10.11.16.20.30.33;
  462,1.10; 465,6.12; 485,14.21.28;
  486,21.32; 487,11.18.20.33
  his double universe, 461,11 ff.,
  487,19 ff.
Anaximander, 452,32; 458,25; 459,1;
  464,18; 465,14; 479,33; 480,1;
  484,12
Anaximenes, 452,31; 458,25; 484,11
Andronicus, 440,14; 450,18
Archytas, 431,12; 467,26; his
  argument for an extracosmic
  space, 467,26 ff.
Aristomenes the Messenian, 470,25
Aristotle, 395,14; 399,28.30; 400,11;
  401,16; 403,3.25; 404,17.26.36;
  405,9.12.17.27; 406,14; 407,5;
  408,28.30.34; 409,9.13; 411,16;
  412,20.32; 413,10;
  414,15.18.21.22.30;
  415,10.12.19.23.32; 420,24;
  421,2.5.7.13.22.32; 422,1.3;
  427,28; 428,2.7; 429,8; 432,16.21;
  433,2.6.13; 438,28; 441,34; 443,30;
  444,3; 451,2.8; 453,28; 454,21;
  455,12.33; 456,5; 458,19; 459,19;
  461,10; 463,3; 464,9; 465,19;
  467,3.5; 468,4.8; 469,2; 472,10;
  476,26; 482,17.20.31; 487,19;
  489,23; 490,14.32; 495,15; 498,34;
  506,18; 507,30; 512,10
Aspasius, 422,20.24; 423,13; 435,13
Athens, 447,25.26.32.33; 448,15.16

change, general account, 394,3-428,19
  change and transformation, 395,25 ff.
  first definition, 413,15
  second definiton, 436,25 ff.
  third definiton, 449,30 ff.
  is in the things changed and
  changers, 402,10 ff.
  Plato on change, 402, 14

Democritus, 395,2; 397,3; 453,2;
  458,26; 459,18; 460,5; 461,28.32;
  462,6.11; 467,16; 512,28
Diares, 517,13.17
Diogenes of Apollonia, 452,32; 458,25;
  475,7
Diogenes of Babylon, 426,2
disagreement between Plato and
  Aristotle, 402,14 ff.
  explained away by Simplicius,
  404,24 ff.
  noted by Proclus, 404,18
  another disagreement, about
  self-changing, 420,14-422,32
discrete receives limitlessness from
  the continuous, 396,30 ff.
division, of the actual into 10
  categories, 400,8 ff.

of the relational, 401,9 ff.
not every division and addition is without limit, 495,33 ff.

Empedoclean, 480,20
Empedocles, 458,30; 459,10.12; 465,6.12
Epicurean, 489,22
Epicurus, 467,2
Euclid, 492,6; 511,31
Eudemus, 411,15; 412,34; 431,6; 433,13; 439,17; 459,25; 466,8; 467,26; 468,5; 474,29; 480,28; 490,11; 493,14; 517,16

(the) great and small, 401,30; 545,7 ff.; 458,10-14; 500,13-16.

heavenly bodies, there is no potentiality in them, 419,7 ff.
Heraclitus, 480,27.33
Hestiaeus, 453,29
Hippasus, 453,13
Hippocrates, 425,2; 461,18

imagination and mathematics, 477,9 ff., 510,28 ff., 517,5 ff.
indefinite dyad, 454,26 ff.
  Alexander's report, 454,26-455,11
indivisible lines, 492,3 ff.

Lacedaemonians, 470,20
Lycia, 404,16
Lyceum, 410,36.37

matter (cannot contain and limit), 502,5 ff.
Melissus, 504,4.5.9

natural bodies, 478,7 ff.
natural places, 483,5 ff.
number, as ideas in Plato, 499,10 ff.
  there is no actually unlimited number, 508,27 ff.

Parmenides, 502,4.6
Peripatetics, 403,6
Plato, 397,15; 401,14; 402,14.15; 404,16.22; 405,4.8.14.16.25; 406,4.13; 420,13.15.18.26; 421,2.3.4.15.18.22.32.33; 428,5; 430,4.12.13.21; 431,1.5.7.8.13; 432,20.34.35; 451,8; 452,29; 453,5.19.26.29.36; 454,20.21.22; 455,8.13.15.17; 458,11; 464,3; 469,2; 471,3; 476,3; 493,17; 498,32; 499,3.9.11.37; 503,11.12; 507,29; 510,28
  his view in *Timaeus* 57E ff., 432,20-9
  his arguments from opposites in the *Phaedo*, 440,35 ff.
  Plato and the Pythagoreans about the causes of change, 430,35 ff.
Platonists, 420,19.20
Plotinus, 403,9; 432,17
Porphyry, 399,19.29; 400,8.10; 407,4; 414,17; 422,2; 428,4; 453,30; 454,17
  on Plato's *Philebus*, 453,30-454,17
potentiality, description of, 400,28-9.
predication of limitlessness, 509,35 ff.
privation is not simply limitlessness, 513,28 ff.
Pythagoras, 428,27; 430,30; 431,1.13; 432,35; 434,14; 452,29; 453,5; 455,12.13.17.20; 456,17; 457,1.17; 458,10; 471,2; 475,8.12.14.22; 476,10.15.18; 499,15
Pythagorean column, 429,9-18

Socrates, 413,23; 464,26
Stoics, 420,11; 480,29

Thales, 452,32; 458,25; 484,11
Thebes, 447,26.27.32; 448,16
Themistius, 400,1; 414,17; 422,20; 428,4; 437,11; 442,1
Theophrastus, 412,35
things that exist in actuality only, 493,9 ff.
time and change, 410,34 ff.

unit (*henas*) no matter, 505,8 ff.
unlimited, difference between Plato and the Pythagoreans, 455,17 ff.
  Pythagorean view, 456,16 ff.
  Anaxagoras, 469,30 ff., 485,20 ff.
  atomists, 460,30 ff.
  and matter, 463,30 ff.
  different senses of the term, 469,29 ff.
  five grounds on which to believe in it, 465,34 ff.
  as untraversable, 470,7 ff.
  belongs to material cause, 514,20 ff.